'A' LEVEL

S

Questions & Answers

by
John S Applegarth BSc

HLT Publications

HLT PUBLICATIONS
200 Greyhound Road, London W14 9RY

First published 1991

ISBN 0 7510 0000 0

British Library Cataloguing-in-Publication.

A CIP Catalogue record for this book is available from the British Library.

Printed and bound in Great Britain.

CONTENTS

PREFACE

This book is intended to complement the HLT Pure Mathematics Textbook for 'A' Level students (and other theoretical textbooks) by selecting questions from previously published examination papers and suggesting methods of solution amplified by advisory notes where appropriate. The material covered is the core common to all examination boards.

It should be noted that examination boards do not publish official solutions to questions they have set nor do they detail their various marking schemes. The suggested solutions are, therefore, the responsibility of the author. The University of Cambridge Local Examinations Syndicate and the University of Oxford Delegacy of Local Examinations bear no responsibility for the example answers to questions taken from their past question papers which are contained in this publication.

The author and publishers gratefully acknowledge the following for their kind permission to reproduce past questions:

Associated Examining Board
University of Cambridge Local Examinations Syndicate
University of Oxford Delegacy of Local Examinations

1 ALGEBRA

1.1 Key formulae and ideas

1. Quadratic function: $ax^2 + bx + c = 0$

   then $$x = \frac{-b \pm \sqrt{(b^2 - 4ac)}}{2a}$$

   a) if $b^2 - 4ac > 0$, solutions are real and different

   $b^2 - 4ac = 0$, solutions are real and equal

   $b^2 - 4ac < 0$, no real solutions

   b) if roots are α and β then

   $$(\alpha + \beta) = -\frac{b}{a} \quad \text{and} \quad \alpha\beta = \frac{c}{a}$$

2. If $f(x)$ eg $(ax^n + bx^{n-1} + cx^{n-2} + \ldots\ldots k)$ can be divided by $(x - \alpha)$ to give a remainder = 0, then $(x - \alpha)$ is a factor of the function.

3. Arithmetic progression: $a,\ a + d,\ a + 2d,\ \ldots\ldots \{a + (n - 1)d\}$

   where a = first term
   d = common difference
   n = number of terms

   Sum to n terms: $S_n = \frac{n}{2}\{2a + (n - 1)d\}$

   Last term: $L = a + (n - 1)d$

   Geometrical progression: $a,\ ar,\ ar^2,\ \ldots\ldots ar^{n-1}$

   where a = first term
   r = common ratio
   n = number of terms

Sum to n terms: $S_n = \dfrac{a(1 - r^n)}{(1 - r)}$

Sum to infinity: $S_\infty = \dfrac{a}{(1 - r)}$

4. Laws of logarithms:

multiplication	$\log_a(x.y)$	$= \log_a(x) + \log_a(y)$
division	$\log_a(x/y)$	$= \log_a(x) - \log_a(y)$
powers	$\log_a(x^n)$	$= n.\log_a(x)$
change of base	$\log_a(x)$	$= \dfrac{\log_b(x)}{\log_b(a)}$

5. Laws of indices:

	$a^x \times a^y$	$= a^{x+y}$
	$a^x \div a^y$	$= a^{x-y}$
if	$a^y = x$	then $y = \log_a x$

1.2 Questions and solutions

Question 1

Express the polynomial

$$f(x) = 2x^4 + x^3 - x^2 + 8x - 4$$

as the product of two linear factors and a quadratic factor $q(x)$.

Prove that there are no real values of x for which $q(x) = 0$.

AEB June 1983

Suggested solution

$$f(x) = 2x^4 + x^3 - x^2 + 8x - 4$$

Try $x = -2$

then $f(2) = 2(-2)^4 + (-2)^3 - (-2)^2 + 8(-2) -4 = 0$

ie there is no remainder,

therefore $(x + 2)$ is a factor of $f(x)$.

To find further factors divide $f(x)$ by $(x + 2)$

$$\begin{array}{r|l} & 2x^3 - 3x^2 + 5x - 2 \\ \hline x + 2 & 2x^4 + x^3 - x^2 + 8x - 4 \\ & 2x^4 + 4x^3 \\ \hline & \quad - 3x^3 - x^2 \\ & \quad - 3x^3 - 6x^2 \\ \hline & \qquad 5x^2 + 8x \\ & \qquad 5x^2 + 10x \\ \hline & \qquad\quad - 2x - 4 \\ & \qquad\quad \underline{- 2x - 4} \end{array}$$

thus two factors are $(x + 2)(2x^3 - 3x^2 + 5x - 2)$

To find the factors of $2x^3 - 3x^2 + 5x - 2$

Try $x = \frac{1}{2}$ $\quad 2\left(\frac{1}{2}\right)^3 - 3\left(\frac{1}{2}\right)^2 + 5\left(\frac{1}{2}\right) - 2 = 0$

thus $\left(x - \frac{1}{2}\right)$ is a factor

To factorise further, divide $2x^3 - 3x^2 + 5x - 2$ by $\left(x - \frac{1}{2}\right)$

$$\begin{array}{r|l} & 2x^2 - 2x + 4 \\ \hline x - \frac{1}{2} & 2x^3 - 3x^2 + 5x - 2 \\ & 2x^3 - x^2 \\ \hline & \quad - 2x^2 + 5x \\ & \quad - 2x^2 + x \\ \hline & \qquad\quad 4x - 2 \\ & \qquad\quad 4x - 2 \end{array}$$

thus factors are $(x + 2)\left(x - \frac{1}{2}\right)(2x^2 - 2x + 4)$

ie $(x + 2)(2x - 1)(x^2 - x + 2)$

where $q(x) = x^2 - x + 2$

To find the roots of the quadratic function $q(x) = 0$

$$x = \frac{-(-1) \pm \sqrt{\{(-1)^2 - 4(1)(2)\}}}{2(1)}$$

ie $(-1)^2 - 4(1)(2) = -7$ which is < 0

There are NO real values for x.

Question 2

Using the factor theorem, or otherwise, find the values of x, in terms of p, which satisfy the equation

$$x^3 - 7p^2x + 6p^3 = 0$$

Hence, or otherwise, find the values of t, in radians to 2 decimal places in the interval $0 \leq t \leq 2\pi$, which satisfy the equation

$$4 \sec^3 t - 7 \sec t + 3 = 0$$

AEB June 1984

Suggested solution

$$x^3 - 7p^2x + 6p^3 = 0$$

Let $x = p$

then $p^3 - 7p^2(p) + 6p^3 = 0$ ie $(x - p)$ is a factor

Divide $x^3 - 7p^2x + 6p^2$ by $(x - p)$ to obtain another factor.

$$\begin{array}{r|l}
 & x^2 + px - 6p^2 \\
\hline
x - p & x^3 \qquad - 7p^2x + 6p^3 \\
 & x^3 - px^2 \\
\cline{2-2}
 & \quad + px^2 - 7p^2x \\
 & \quad + px^2 - p^2x \\
\cline{2-2}
 & \qquad - 6p^2x + 6p^3 \\
 & \qquad - 6p^2x + 6p^3 \\
\cline{2-2}
 & \qquad \quad 0 \quad 0
\end{array}$$

ie there is no remainder, therefore the factors of $x^3 - 7p^2x + 6p^2$ are $(x - p)(x^2 + px - 6p^2)$.

However, the second bracket can be factorised further

ie $\quad x^2 + px - 6p^2 = (x + 3p)(x - 2p)$

thus the factors of $x^3 - 7p^2x + 6p^3$ are $(x - p)(x - 2p)(x + 3p) = 0$

ie $\quad x$ has values of $+p$, $+2p$, $-3p$

$$4\sec^3 t - 7\sec t + 3 = 0$$

Compare this equation with the cubic equation in x above

ie $\quad \sec^3 t - 7\left(\frac{1}{2}\right)^2 \sec t + 3\left(\frac{1}{2}\right)^2 = 0$

or $\quad \sec^3 t - 7\left(\frac{1}{2}\right)^2 \sec t + 6\left(\frac{1}{2}\right)^3 = 0$

by comparison $x = \sec t$ and $p = \frac{1}{2}$

and values of sec t are $\frac{1}{2}$ OR $2\left(\frac{1}{2}\right)$ OR $-3\left(\frac{1}{2}\right)$

solutions:

i) $\sec t = \frac{1}{2}$ ie $\cos t = 2$ ... not possible

ii) $\sec t = 1$ ie $\cos t = 1 \therefore t = 0$ rads or $t = 2\pi$ rads

iii) $\sec t = \frac{-3}{2}$ ie $\cos t = \frac{-2}{3} \therefore t = 2{\cdot}30$ rad or $t = 3{\cdot}98$ rad

Question 3

The roots of the equation $2x^2 + 5x + 7 = 0$ are α and β. Without solving the equation, find the value of $\alpha^2 + \beta^2$.

What can you deduce about the roots α and β?

Show that $8(\alpha^3 + \beta^3) = 85$ and find a quadratic equation with roots $\frac{\alpha^2}{\beta}$ and $\frac{\beta^2}{\alpha}$

Oxford Summer 1988

Suggested solution

$$2x^2 + 5x = 7 = 0$$

If the roots are α and β then

$$\alpha + \beta = -\frac{5}{2} \qquad \text{and} \qquad \alpha\beta = \frac{7}{2}$$

but $\alpha^2 + \beta^2$

$$\begin{aligned} &= \alpha^2 + 2\alpha\beta + \beta^2 - 2\alpha\beta \\ &= (\alpha + \beta)^2 - 2\alpha\beta \\ &= \left\{\frac{-5}{2}\right\}^2 - 2\left\{\frac{7}{2}\right\} = \frac{25}{4} - 7 \\ &= \frac{-3}{4} \end{aligned}$$

α^2 and β^2 must both be positive values but $\alpha^2 + \beta^2 = \frac{-3}{4}$

$\therefore$ α and β are complex numbers.

Factorise $8(\alpha^3 + \beta^3)$

$$\begin{aligned} &= 8(\alpha + \beta)(\alpha^2 - \alpha\beta + \beta^2) \\ &= 8\left(\frac{-5}{2}\right)\left(\frac{-3}{4} - \frac{7}{2}\right) \\ &= +85 \end{aligned}$$

Let new equation be: $Ax^2 + Bx + C = 0$

Sum of roots $\frac{\alpha^2}{\beta} + \frac{\beta^2}{\alpha} = -\frac{B}{A}$

$\therefore \qquad \dfrac{\alpha^3 + \beta^3}{\alpha\beta} = -\dfrac{B}{A}$

$$\frac{\frac{85}{8}}{\frac{7}{2}} = \frac{85}{28} = -\frac{B}{A}$$

Product of roots $\qquad \dfrac{\alpha^2}{\beta} \cdot \dfrac{\beta^2}{\alpha} = \dfrac{C}{A}$

ie $\qquad \alpha\beta = \dfrac{7}{2} = \dfrac{C}{A}$

that is $\qquad \dfrac{C}{A} = \dfrac{98}{28}$

hence new equation is: $\qquad x^2 + \dfrac{B}{A}x + \dfrac{C}{A} = 0$

ie $\qquad x^2 - \dfrac{85}{28}x + \dfrac{98}{28} = 0$

ie $\qquad 28x^2 - 85x + 98 = 0$

Question 4

The quadratic function f is defined by

$f : x \rightarrow 2x^2 + 4x + 5$, where $x \varepsilon$ R.

a) Find the set of values of x for which $f(x) > 3x^2$.

b) Find the set of values of k for which the equation $f(x) = kx$ has no real roots.

c) By considering the identity

$$2x^2 + 4x + 5 = 2(x + A)^2 + B$$

where A and B are constants, find the greatest value of $\frac{1}{f(x)}$.

d) The roots of the equation f(x) = 0 are α and β.

Find an equation with numerical coefficients whose roots are $\alpha^2 + 2$ and $\beta^2 + 2$.

AEB June 1986

Suggested solution

a) $f : x = 2x^2 + 4x + 5 > 3x^2$

ie $-x^2 + 4x + 5 > 0$

$-(x + 1)(x - 5) > 0$

$(x + 1)(-x + 5) > 0$

value of: x	$(x + 1)$	$(-x + 5)$	$(x + 1)(-x + 5)$
< –1	–ve	+ve	–ve
= –1	0	+ve	0
> –1	+ve	+ve	+ve
= 0	+ve	+ve	+ve
< +5	+ve	+ve	+ve
= +5	+ve	0	0
> +5	+ve	–ve	–ve

Thus set of values of x: $-1 < x < +5$

---

b) $2x^2 + 4x + 5 = kx$

$2x^2 + (4-k)x + 5 = 0$

hence $x = \dfrac{-(4-k) \pm \sqrt{(4-k)^2 - 4(2)(5)}}{2(2)}$

there are no real roots if $(4-k)^2 - 40 < 0$

ie $k^2 - 8k - 24 < 0$

limits of k are given by

$$k = \frac{-(-8) \pm \sqrt{(-8)^2 - 4(1)(-24)}}{2(1)}$$

$$k = 4 \pm 2\sqrt{10}$$

$\therefore$ limits of k are $4 - 2\sqrt{10} < k < 4 + 2\sqrt{10}$

---

c) $2x^2 + 4x + 5 = 2(x + A)^2 + B$

$= 2(x^2 + 2Ax + A^2) + B$

$= 2x^2 + 4Ax + (2A^2 + B)$

equate coefficients $4A = 4 \quad \therefore \underline{A = 1}$

$2A^2 + B = 5$

$2 + B = 5 \quad \therefore \underline{B = 3}$

hence $2x^2 + 4x + 5 = 2(x + 1)^2 + 3$

thus $\dfrac{1}{f(x)} = \dfrac{1}{2(x + 1)^2 + 3}$

greatest value when $(x + 1)^2 = 0$ ie $x = -1$

Maximum $\dfrac{1}{f(x)} = \dfrac{1}{2(0)^2 + 3} = \dfrac{1}{3}$

---

d) Roots of $2x^2 + 4x + 5$ are α and β.

hence α and $\beta = -\dfrac{(4)}{2} = -2$ and $\alpha\beta = \dfrac{5}{2}$

Let new roots be A, B where $A = \alpha^2 + 2$ and $B = \beta^2 + 2$

hence $A + B = (\alpha^2 + 2) + (\beta^2 + 2) = \alpha^2 + \beta^2 + 4$

$= (\alpha + \beta)^2 - 2\alpha\beta + 4$

$= (-2)^2 - 2\dfrac{(5)}{2} + 4 = \underline{3}$

and AB $= (\alpha^2 + 2)(\beta^2 + 2) = \alpha^2\beta^2 + 2(\alpha^2 + \beta^2) + 4$

$$= \left(\frac{5}{2}\right)^2 + 2(-2)^2 - (2)(2)\left(\frac{5}{2}\right) + 4$$

$$= \frac{33}{4}$$

New equation: $x^2 + \frac{b}{a}x + \frac{c}{a} = 0$

ie $x^2 - 3x + \frac{33}{4} = 0$

$$4x^2 - 12x + 33 = 0$$

Question 5

a) An arithmetic progression is such that the sum of the first 12 terms is 270 and the sum of the first 17 terms is 510. Find the first term and the common difference.

b) A geometric progression is such that the sum of the first two terms is 1, and the sum of the first four terms is 5. Given that all the terms are positive, find the first term and the common ratio.

c) A new series is formed as follows. The kth term of the new series is the product of the kth term of the arithmetic progression in (a) and the kth term of the geometric progression in (b).

Show that the kth term of the new series is $(k + 1)2^{k-1}$.

Prove by induction that

$$\sum_{k=1}^{n} (k + 1)2^{k-1} = n\,2^n$$

Oxford June 1988

Suggested solution

a) Arithmetic progression is a, a+d, a+2d, ... a + (n–1)d

where a = first term,

d = common difference,

n = number of terms

Sum to n terms = $S_n = \frac{n}{2}\{2a + (n-1)d\}$

Given $S_{12} = \frac{12}{2}\{2a + (12-1)d\} = 270$

ie $2a + 11d = 45$ ...... (i)

also $S_{17} = \frac{17}{2}\{2a + (17-1)d\} = 510$

$a + 8d = 30$ ..... (ii)

multiply equation (ii) by 2, then subtract equation (i)

ie $2a + 16d = 60$

$5d = 15$

$\underline{d = 3}$

substitute d = 3 into equation (ii)

$$a + 8(3) = 30$$

$$\underline{a = 6}$$

b) Geometric progression is $a, ar, ar^2, ar^3 \ldots\ldots ar^{n-1}$

where a = first term, r = common ratio

n = number of terms

Sum to n terms = $S_n = \dfrac{a(1 - r^n)}{(1 - r)}$

Given $S_2 = \dfrac{a(1 - r^2)}{(1 - r)} = 1$

ie $a(1 + r) = 1 \quad \ldots\ldots \quad \text{(iii)}$

also $S_4 = \dfrac{a(1 - r^4)}{(1 - r)} = 5$

ie $\dfrac{a(1 - r^2)(1 + r^2)}{(1 - r)} = 5$

$a(1 + r)(1 + r^2) = 5 \quad \ldots\ldots \quad \text{(iv)}$

divide equation (iv) by equation (iii)

ie $(1 + r^2) = 5 \qquad r^2 = 4 \qquad r = \pm 2$

given that all terms are positive, if $r = -2$ then odd powered terms would be negative

$\therefore \qquad \underline{r = +2}$

substitute in equation (iii)

$$a(1 + 2) = 1$$

$$a = \frac{1}{3}$$

c) k^{th} term of the arithmetic progression $= 6 + (k - 1)3$

$= 3(k + 1)$

k^{th} term of the geometrical progression $= \frac{1}{3}(2)^{k-1}$

k^{th} term of the new series $= 3(k+1) \cdot \frac{1}{3}(2)^{k-1} = 2^{k-1}(k+1)$

By induction: $\sum_{k=1}^{n} (k+1)2^{k-1} = k.2^k$

ie $(1+1)2^{1-1} + (2+1)2^{2-1} + (3+1)2^{3-1} \ldots\ldots (k+1)2^{k-1} = k.2^k$

Assume that this is true for n = k; add the next term to each side

$$2.2^0 + 3.2^1 + 4.2^2 \ldots\ldots + (k+1)2^{k-1} + (k+1+1)2^{k+1-1}$$

$$= k.2^k + (k+1+1)2^{k+1-1}$$

$$= k.2^k + (k+2)2^k$$

$$= 2^k(k+k+2) = 2^k.2(k+1)$$

$$= 2^{k+1}(k+1)$$

but $n = k+1$ $\qquad = n.2^n$

hence $\sum_{k=1}^{n} (k+1)2^{k-1} = n.2^n$

Question 6

An infinite geometric series has first term a and common ratio r, where $a > 0$ and $0 < r < 1$. Show that D, the difference between the sum to infinity of the series and the sum of the first n terms of the series, is given by

$$D = \frac{ar^n}{1 - r}$$

Deduce that $\log_{10}D = n.\log_{10}r + \log_{10}\left[\frac{a}{1 - r}\right]$

The following table gives values of D, rounded to the nearest 100, corresponding to some values of n.

n	4	6	10	14	16
D	4200	3000	1600	800	600

By drawing an appropriate linear graph, estimate, to 2 significant figures, a value for r and a value for a.

AEB June 1986

Suggested solution

Geometric series: $a, ar, ar^2, \quad ..... \quad ar^n$

Sum of n terms $= S_n = \frac{a(1 - r^n)}{(1 - r)}$

Sum to infinity $= S_\infty = \frac{a}{(1 - r)}$ when $r < 1$

$\therefore \quad D = S_\infty - S_n = \frac{a}{(1 - r)} - \frac{a(1 - r^n)}{(1 - r)} = \frac{ar^n}{(1 - r)}$

hence $\log_{10}D = \log_{10}(r^n) + \log_{10}\left\{\frac{a}{1 - r}\right\}$

ie $\log_{10}D = n.\log_{10}(r) + \log_{10}\left\{\frac{a}{1 - r}\right\}$

Compare this equation with the standard equation of a straight line $y = mx + c$ where m is the gradient and c is the intercept on the vertical y axis.

Here y relates to $\log_{10}D$; m relates to $\log_{10}(r)$; x relates to n and c to $\log_{10}\left\{\frac{a}{1 - r}\right\}$

Plotting of the graph:

n	4	6	10	14	16
D	4200	3000	1600	800	600
$\log_{10}D$	3.623	3.477	3.204	2.903	2.778

Plot values of $\log_{10}D$ on the vertical axis and values of n on the horizontal axis. The scales on the axes (which will not be the same) should be chosen to make as much use of the graph paper as possible. In the example shown the vertical scale for $\log_{10}D$ is 5cm to represent 1.000 and the horizontal scale for n is 1cm to represent 1.

Measurements from the graph:

i) Gradient ... construct a right angled triangle making measurements over as large distances as is possible but within the confines of the points actually plotted.

$$\text{gradient} = -\frac{[3.920 - 2.840]}{[15 - 0]} = -\frac{1.08}{15} = -0.072$$

ie $\log_{10} r = -0.072$ $\quad\therefore r = 0.85$

ii) Intercept on $\log_{10}D$ axis $= 3.920 = \log_{10}\dfrac{a}{(1 - r)}$

$$\text{ie } 8318 = \frac{a}{(1 - 0.85)} = \frac{a}{0.15}$$

hence $a = 1248 = 1200$ to 2 sig fig

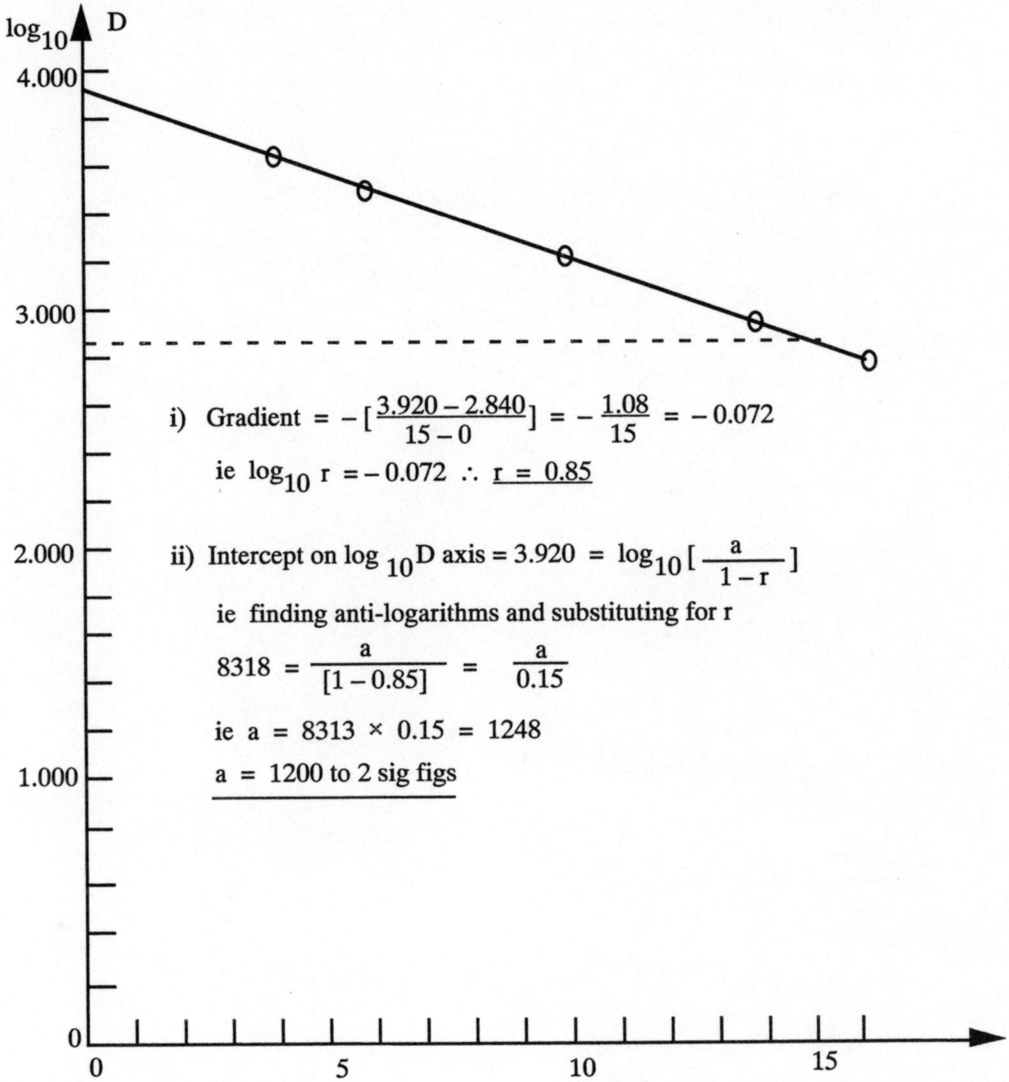

i) Gradient $= -\left[\frac{3.920 - 2.840}{15 - 0}\right] = -\frac{1.08}{15} = -0.072$

ie $\log_{10} r = -0.072 \therefore \underline{r = 0.85}$

ii) Intercept on $\log_{10} D$ axis $= 3.920 = \log_{10}\left[\frac{a}{1-r}\right]$

ie finding anti-logarithms and substituting for r

$8318 = \frac{a}{[1 - 0.85]} = \frac{a}{0.15}$

ie $a = 8313 \times 0.15 = 1248$

$\underline{a = 1200 \text{ to 2 sig figs}}$

2 POWER SERIES

2.1 Key formulae and ideas

1. Pascal's triangle

... used to determine coefficients in expansions of

$(a + b)^n \Rightarrow a^n b^0 + a^{n-1} b^1 + a^{n-2} b^2 + \ldots\ldots a^1 b^{n-1} + a^0 b^n$

				1						
			1		1					
		1		2		1				$(a + b)^2$
	1		3		3		1			$(a + b)^3$
1		4		6		4		1		$(a + b)^4$

and so on ...

2. Factorial n → $n! = n(n-1)(n-2)(n-3) \ldots\ldots\ldots\ldots (1)$

Special case $O! = 1$

3. Binomial coefficients are given by

$$^nC_r \text{ or } \begin{bmatrix} n \\ r \end{bmatrix} = \frac{n(n-1)(n-2)(n-3) \ldots\ldots (n-r+1)}{r!}$$

$$= \frac{n!}{r!\,(n-r)!} = \begin{bmatrix} n \\ n-r \end{bmatrix}$$

4. Exponential series

$$e^x = 1 + x + \frac{x^2}{2!} + \frac{x^3}{3!} + \ldots\ldots \qquad \text{where } -\infty < x < \infty$$

5. Logarithmic series

$$\log_e (1 + x) = \mathrm{l_n}x = x - \frac{x^2}{2} + \frac{x^3}{3} - \frac{x^4}{4} + \ldots \text{ in the range } -1 < x \leq 1$$

$$\log_e (1 - x) = -x - \frac{x^2}{2} - \frac{x^3}{3} - \frac{x^4}{4} - \ldots \qquad \text{in the range } -1 \leq x < 1$$

6. Trigonometrical series

$$\sin x = x - \frac{x^3}{3!} + \frac{x^5}{5!} - \frac{x^7}{7!} + \ldots \qquad \text{in the range } -\infty < x < \infty$$

$$\cos x = 1 - \frac{x^2}{2!} + \frac{x^4}{4!} - \frac{x^6}{6!} + \ldots \qquad \text{in the range } -\infty < x < \infty$$

7. Maclaurin series

$$\mathrm{f}(x) = a_0 + a_1x + a_2x^2 + a_3x^3 + \ldots$$

$x = 0$ $$\mathrm{f}(0) = a_0 + a_1(0) + a_2(0) + \ldots = a^0$$

$$\mathrm{f}'(x) = a_1 + 2a_2x + 3a_3x^2 + \ldots$$

$x = 0$ $$\mathrm{f}'(0) = a_1 + 2a_2(0) + 3a_3(0) + \ldots = a_1 = 1!a_1$$

$$\mathrm{f}''(x) = 2a_2 + 6a_3x + 12a_4x^2 + \ldots$$

$x = 0$ $$\mathrm{f}''(0) = 2a_2 + 6a_3(0) + 12a_4(0)^2 + \ldots = 2a_2 = 2!a_2$$

etc ...

To find coefficients a_1, a_2, a_3 etc use successive differentiation and put $x = 0$

ie $$f(x) = f(0) + f'(0)x + \frac{f''(0)}{2!}x^2 + \frac{f'''(0)}{3!}x^3 + \ldots$$

2.2 Questions and solutions

Question 1

Given that $|x| < 1$, find in ascending powers of x, up to and including the terms in x^3, series expansions in which the coefficients are simplified for

a) $(1+x)^{\frac{1}{2}}$

b) $\left[1 + \frac{x}{4}\right]^{-1}$

c) Prove that if x were sufficiently small for terms in x^4 and higher powers of x to be neglected, then

$$(1+x)^{\frac{1}{2}} - \frac{4+3x}{4+x} = -\frac{1}{32}x^3$$

AEB June 1984

Suggested solution

a) $$(1+x)^{\frac{1}{2}} = 1 + \frac{1}{2}(x) + \frac{\left(\frac{1}{2}\right)\left(-\frac{1}{2}\right)(x^2)}{2!} + \frac{\left(\frac{1}{2}\right)\left(-\frac{1}{2}\right)\left(-\frac{3}{2}\right)}{3!}x^3$$

$$= 1 + \frac{x}{2} - \frac{x^2}{8} + \frac{x^3}{16}$$

b) $$\left[1 + \frac{x}{4}\right]^{-1} = 1 + (-1)\left[\frac{x}{4}\right] + \frac{(-1)(-2)}{2!}\left[\frac{x}{4}\right]^2 + \frac{(-1)(-2)(-3)}{3!}\left[\frac{x}{4}\right]^3$$

$$= 1 - \frac{x}{4} + \frac{x^2}{16} - \frac{x^3}{64}$$

c) $$\frac{(4+3x)}{(4+x)} = (4+3x)(4+x)^{-1} = \frac{1}{4}(4+3x)\left[1 + \frac{x}{4}\right]^{-1}$$

$$= \frac{1}{4}(4+3x)\left[1 - \frac{x}{4} + \frac{x^2}{16} - \frac{x^3}{64}\right]$$

$$= \frac{1}{4}\left[4 - x + \frac{x^2}{4} - \frac{x^3}{16} + 3x - \frac{3x^2}{4} + \frac{3x^3}{16}\right]$$

$$= 1 + \frac{x}{2} - \frac{x^2}{8} + \frac{x^3}{32}$$

hence combine this result with that of (a) above:

$$(1+x)^{\frac{1}{2}} - \left[\frac{4+3x}{3+x}\right] = 1 + \frac{x}{2} - \frac{x^2}{8} + \frac{x^3}{16} - 1 - \frac{x}{2} + \frac{x^2}{8} - \frac{x^3}{32}$$

$$= \frac{x^3}{32}$$

Question 2

Write down the first four terms of the binomial series expansion in ascending powers of x of the function $(1-x)^{-\frac{1}{2}}$, and hence show that

$$\sqrt{10} = 3\left[1 + \frac{1}{20} + \frac{1 \times 3}{20 \times 40} + \frac{1 \times 3 \times 5}{20 \times 40 \times 60}\right]$$

Oxford June 1982

Suggested solution

General binomial expansion is given by

$$(a + x)^n = a^n + na^{n-1}x + \frac{n(n-1)a^{n-2}}{2!}x^2 + \frac{n(n-1)(n-2)a^{n-3}}{3!}x^3 + \ldots \text{ etc}$$

In this case put $n = -\frac{1}{2}$ and $x = -x$

$$(1-x)^{-\frac{1}{2}} = 1^{-\frac{1}{2}} + \left(-\frac{1}{2}\right)(1)^{-\frac{3}{2}}(-x) + \frac{\left(-\frac{1}{2}\right)\left(-\frac{3}{2}\right)(1)^{-\frac{5}{2}}}{2!}(-x)^2$$

$$+ \frac{\left(-\frac{1}{2}\right)\left(-\frac{3}{2}\right)\left(-\frac{5}{2}\right)(1)^{-\frac{7}{2}}}{3!}(-x)^3$$

$$= 1 + \frac{x}{2} + \frac{3}{8}x^2 + \frac{5}{16}x^3$$

$$\sqrt{10} = (1+9)^{\frac{1}{2}} = 9^{\frac{1}{2}}\left[\frac{1}{9} + 1\right]^{\frac{1}{2}} = 3\left[1 + \frac{1}{9}\right]^{\frac{1}{2}}$$

Put $\left[1 + \frac{1}{9}\right]^{\frac{1}{2}} = [1-x]^{-\frac{1}{2}}$

ie $$\frac{10}{9} = \frac{1}{[1-x]} \quad \therefore x = \frac{1}{10}$$

thus $$\sqrt{10} = 3\left[1 - \frac{1}{10}\right]^{-\frac{1}{2}}$$

$$= 3\left[1 + \frac{1}{2} \bullet \frac{1}{10} + \frac{3}{8} \bullet \frac{1^2}{10^2} + \frac{5}{16} \bullet \frac{1^3}{10^3}\right]$$

$$= 3\left[1 + \frac{1}{20} + \frac{1 \text{ x } 3}{20 \text{ x } 40} + \frac{5}{20 \text{ x } 40 \text{ x } 20}\right]$$

ie $$\sqrt{10} = 3\left[1 + \frac{1}{20} + \frac{1 \text{ x } 3}{20 \text{ x } 40} + \frac{1 \text{ x } 3 \text{ x } 5}{20 \text{ x } 40 \text{ x } 60}\right]$$

Question 3

Use the series expansion for sin x to write down in ascending powers of x the first three non-zero terms in the series expansion of $\frac{\sin x}{x}$.

Write down in ascending powers of y the first three non-zero term in the series expansion of $\log_e(1 - y)$.

Use these two series to show that if x is small enough for terms in x^6 and higher powers of x to be neglected, then

$$\log_e(\sin x) - \log_e(x) = Ax^2 + Bx^4$$

for constants A and B whose numerical values are to be found.

AEB June 1983

Suggested solution

The series expansion of sin x is:

$$\sin x = x - \frac{x^3}{3!} + \frac{x^5}{5!} - \frac{x^7}{7!} + \ldots\ldots \frac{(-1)^{n-1}x^{2n-1}}{(2n-1)!}$$

The first three non-zero terms of $\frac{\sin x}{x}$ are:

$$\frac{\sin x}{x} = 1 - \frac{x^2}{6} + \frac{x^4}{120} - \frac{x^6}{5040}$$

The series expansion of $\log_e(1 + x)$ is:

$$\log_e(1 + x) = x - \frac{x^2}{2} + \frac{x^3}{3} - \frac{x^4}{4} + \ldots\ldots \frac{(-1)^{n-1}x^n}{n}$$

Put $x = -y$, then the first three non-zero terms are:

$$\log_e (1 - y) = -y - \frac{y^2}{2} - \frac{y^3}{3}$$

From above $\quad \sin x = x - \frac{x^3}{3!} + \frac{x^5}{5!} - .....$

$$= x\left[1 - \frac{x^2}{3!} + \frac{x^4}{5!} - .....\right]$$

Write $\quad \sin x = x(1 - y)$ where $y = \frac{x^2}{3!} - \frac{x^4}{5!}$

then $\quad \log_e (\sin x) = \log_e (x) + \log_e (1 - y)$

ie $\quad \log_e (\sin x) - \log_e (x) = \log_e (1 - y)$

and $\quad \log_e (1 - y) = -\left[\frac{x^2}{3!} - \frac{x^4}{5!}\right] - \left[\frac{x^2}{3!} - \frac{x^4}{5!}\right]^2 - \left[\frac{x^2}{3!} - \frac{x^4}{5!}\right]^3$

ie $\quad \log_e (1 - y) = -\frac{x^2}{6} - \frac{x^4}{120} - \left[\frac{x^4}{36} - \frac{2x^6}{720} \cdots\right] - \left[\frac{x^6}{216} \cdots\right]$

and neglecting terms in x^6 and above

$$= -\frac{x^2}{6} + \frac{[3x^4 - 10x^4]}{360}$$

$$= -\frac{1}{6}(x^2) - \frac{7}{360}(x^4)$$

which is of the form $Ax^2 + Bx^4$

where $A = -\frac{1}{6}$ and $B = -\frac{7}{360}$

Question 4

Prove that $\frac{d}{dx}\tan^{-1}x = \frac{1}{1+x^2}$

Obtain Maclaurin's expansions in ascending powers of x up to and including the term x^3 for each of the functions

a) $\tan x$

b) $\tan^{-1} x$

Hence or otherwise find the limit, as x tends to zero, of $\frac{\tan x - \tan^{-1} x}{x^3}$

Oxford June 1987

Suggested solution

i) Let $z = \tan^{-1}x$ $\quad\therefore\quad$ $x = \tan z$ and $\frac{dx}{dz} = \sec^2 z$

thus $\frac{dz}{dx} = \frac{1}{\sec^2 z} = \frac{1}{(\tan^2 z + 1)} = \frac{1}{1+x^2}$

ii) By Maclaurin's theorem:

$$f(x) = f(0) + \frac{x}{1!}f'(0) + \frac{x^2}{2!}f''(0) + \frac{x^3}{3!}f'''(0) +$$

a) Let $f(x) = \tan x$ $\qquad f(0) = 0$

$f'(x) = \sec^2 x$ $\qquad f'(0) = 1$

$f''(x) = 2\sec x\{\sec x.\tan x\}$

$= 2\sec^2 x \,.\, \tan x$

$= 2\{\tan^3 x + \tan x\}$ $\qquad f''(0) = 0$

$f'''(x) = 2\{3\tan^2 x \,.\, \sec^2 x + \sec^2 x\}$

$$= 2\sec^2 x\,\{3\tan^2 x + 1\} \qquad f'''(0) = 2$$

Substituting in the expression for f(x)

$$\tan x = 0 + \frac{x}{1}(1) + 0 + \frac{x^3}{3!}(2)$$

ie
$$\tan x = x + \frac{x^3}{3}$$

b) Let

$$f(x) = \tan^{-1} x \qquad f(0) = 0$$

$$f'(x) = \frac{1}{(1 + x^2)} = (1 + x^2)^{-1} \qquad f'(0) = 1$$

$$f''(x) = -1\,\{1 + x^2\}^{-2}\,\{2x\} \qquad f''(0) = 0$$

$$f'''(x) = -1\,\{2(1 + x^2)^{-2} + 2x(-2)(1 + x^2)^{-3}\,(2x)\}$$

$$= -\frac{2}{(1 + x^2)^2} + \frac{6x^2}{(1 + x^2)^3} \qquad f'''(0) = -2$$

Substituting in the expression for f(x)

$$\tan^{-1} x = 0 + x(1) + \frac{x^2}{2!}(0) + \frac{x^3}{3!}(-2)$$

$$\tan^{-1} x = x - \frac{x^3}{3}$$

iii) Hence

$$\frac{\tan x - \tan^{-1} x}{x^3} = \frac{\left[x + \frac{x^3}{3}\right] - \left[x - \frac{x^3}{3}\right]}{x^3}$$

$$= \frac{2x^3}{3x^3} = \frac{2}{3}$$

3 TRIGONOMETRY

3.1 Key formulae and ideas

1. $\sin^2 A + \cos^2 A = 1$

   $\sin^2 A = 2.\sin A.\cos A$

   $\cos^2 A = \cos^2 A - \sin^2 A = 1 - 2.\sin^2 A = 2.\cos^2 A - 1$

2. Compound angles

   $\sin(A \pm B) = \sin A.\cos B \pm \cos A.\sin B$

   $\cos(A \pm B) = \cos A.\cos B \mp \sin A.\sin B$

   $$\tan(A \pm B) = \frac{\tan A \pm \tan B}{1 \mp \tan A.\tan B}$$

3. Functions of angles in all quadrants based on quadrant 1

	$-A$	$90^\circ \pm A$ $\frac{\pi}{2} \pm A$	$180^\circ \pm A$ $\pi \pm A$	$270^\circ \pm A$ $\frac{3\pi}{2} \pm A$	$360^\circ \pm A$ $2\pi \pm A$
sin	$-\sin A$	$\cos A$	$\mp \sin A$	$-\cos A$	$\pm \sin A$
cos	$\cos A$	$\mp \sin A$	$-\cos A$	$\pm \sin A$	$\cos A$
tan	$-\tan A$	$\mp \cot A$	$\pm \tan A$	$\mp \cot A$	$\pm \tan A$

4 Sum, difference and product of functions of angles

$$\sin A + \sin B = 2\sin\left[\frac{A+B}{2}\right]\cos\left[\frac{A-B}{2}\right]$$

$$\sin A - \sin B = 2\cos\left[\frac{A+B}{2}\right]\sin\left[\frac{A-B}{2}\right]$$

$$\cos A + \cos B = 2\cos\left[\frac{A+B}{2}\right]\cos\left[\frac{A-B}{2}\right]$$

$$\cos A - \cos B = 2\sin\left[\frac{A+B}{2}\right]\sin\left[\frac{A-B}{2}\right]$$

$$\sin A.\sin B = \frac{1}{2}\{\cos(A-B) - \cos(A+B)\}$$

$$\cos A.\cos B = \frac{1}{2}\{\cos(A-B) + \cos(A+B)\}$$

$$\sin A.\cos B = \frac{1}{2}\{\sin(A-B) + \sin(A+B)\}$$

3.2 Questions and solutions

Question 1

Express cos x + sin x in the form R.cos($x - \alpha$), where R ≥ 0 and $0 \leq \alpha < 2\pi$, giving the value of R and the value of α.

i) Sketch the graph of y = cos x + sin x for $-2\pi \leq x \leq 2\pi$

ii) By drawing appropriate lines on your graph, determine the number of roots lying in the interval $-2\pi \leq x \leq 2\pi$, of each of the following equations:

a) $\cos x + \sin x = -\dfrac{1}{2}$

b) $\cos x + \sin x = 2$

iii) Find the set of values of x, lying in the interval $-\pi \leq x \leq \pi$, for which

$|\cos x + \sin x| > 1.$

Cambridge November 1986

Suggested solution

Let $\cos x + \sin x = R.\cos(x - \alpha)$

$= R\{\cos x.\cos\alpha + \sin x.\sin\alpha\}$

then $R.\cos\alpha = 1$

and $R.\sin\alpha = 1$

$$\tan\alpha = \frac{R.\sin\alpha}{R.\cos\alpha} = 1$$

ie $\alpha = 45^o$ and $\cos\alpha = \sin\alpha = \dfrac{1}{\sqrt{2}}$

thus $R\left[\dfrac{1}{\sqrt{2}}\right] = 1$ $\therefore R = \sqrt{2}$

ie $\cos x + \sin x = \sqrt{2}\cos(x - 45)^o$

i) A sketch graph is not required to be drawn to scale on graph paper but should look realistic in terms of scale intervals and be drawn neatly. Critical values for the construction of the sketch should be calculated.

ii) In the range $-2\pi < x < 2\pi$

$x = -2\pi \qquad y = \cos(-2\pi) + \sin(-2\pi) \qquad = 0 + 1 = 1$

$x = \frac{7\pi}{4} \qquad y = \cos\left[\frac{-7\pi}{4}\right] + \sin\left[\frac{-7\pi}{4}\right] \qquad = 0.7071 + 0.7071 = 1.4142$

$x = -\frac{3\pi}{2} \qquad y = \cos\left[\frac{-3\pi}{2}\right] + \sin\left[\frac{-3\pi}{2}\right] \qquad = 0 + 1 = 1$

$x = \frac{-5\pi}{4} \qquad y = \cos\left[\frac{-5\pi}{4}\right] + \sin\left[\frac{-5\pi}{4}\right] \qquad = -0.7071 + 0.7071 = 0$

$x = -\pi \qquad y = \cos[-\pi] + \sin[-\pi] \qquad = -1 + 0 = -1$

$x = -\frac{3\pi}{4} \qquad y = \cos\left[\frac{-3\pi}{4}\right] + \sin\left[\frac{-3\pi}{4}\right] \qquad = -0.7071 - 0.7071 = -1.4142$

$x = -\frac{\pi}{2} \qquad y = \cos\left[\frac{-\pi}{2}\right] + \sin\left[\frac{-3\pi}{2}\right] \qquad = 0 - 1 = -1$

$x = -\frac{\pi}{4} \qquad y = \cos\left[\frac{-\pi}{4}\right] + \sin\left[\frac{-\pi}{4}\right] \qquad = 0.7071 - 0.7071 = 0$

$x = 0 \qquad y = \cos[0] + \sin[0] \qquad = 1 + 0 = 1$

The values for y are now repetitive from 0 to 2π

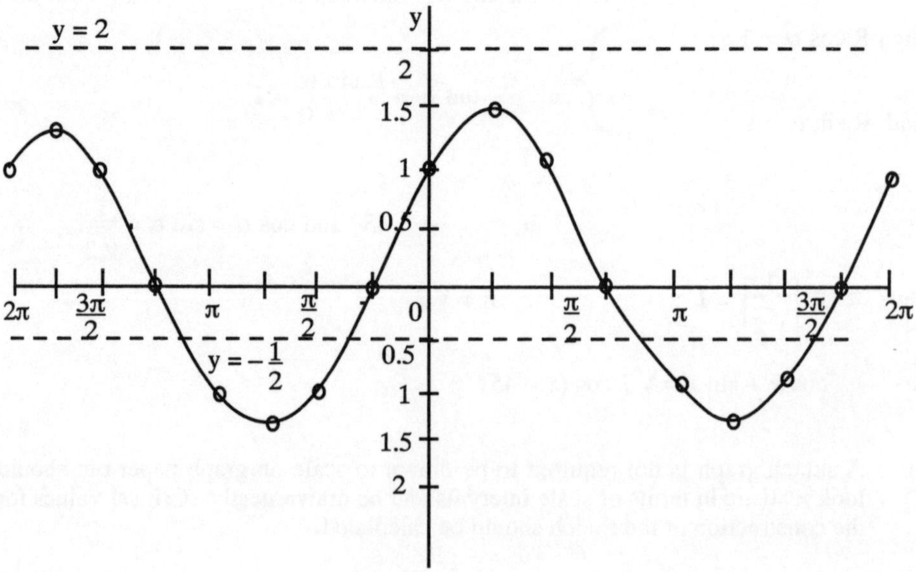

a) $\cos x + \sin x = -\frac{1}{2}$ ie $y = -\frac{1}{2}$, a constant value

Draw a straight line parallel to the x–axis at the value $y = -\frac{1}{2}$. The line cuts the curve $y = \cos x + \sin x$ in 4 places between the values $-2\pi \le x \le 2\pi$.

ie the number of roots = 4

Note: The question does not require the determination of the value of the 4 roots.

b) $\cos x + \sin x = 2$ ie $y = 2$, a constant value.

Again draw a straight line parallel to the x–axis at the value $y = 2$. The line does not cut the curve which has a maximum value of 1.4142

ie there are NO REAL roots for $\cos x + \sin x = 2$.

Note: This can be checked by forming a quadratic equation in either $\sin x$ or $\cos x$ in which case it can be seen that the term $b^2 < 4ac$ requires the square root of a negative number.

iii) $|\cos x + \sin x| > 1$ in the range $-\pi \le x \le \pi$

either $\cos x + \sin x > +1$ which is satisfied by $0 < x < \frac{\pi}{2}$

or $\cos x + \sin x < -1$ which is satisfied by $-\pi < x < \frac{-\pi}{2}$

Question 2

Express $\sqrt{6}.\cos\theta - \sqrt{2}.\sin\theta$ in the form $R\cos(\theta+\alpha)$

where $R > 0$ and $0 < \alpha < \frac{1}{2}\pi$

Find the general solution, in radians, of the equation $(\sqrt{6}.\cos\theta - \sqrt{2}.\sin\theta)^2 = 4$

Oxford June 1988

Suggested solution

$$R\cos(\theta + \alpha) = R\{\cos\theta.\cos\alpha - \sin\theta.\sin\alpha\}$$

Now $\sqrt{6}.\cos\theta - \sqrt{2}.\sin\theta = \sqrt{2}\{\sqrt{3}.\cos\theta - 1.\sin\theta\}$

$$= 2\sqrt{2}\left[\frac{\sqrt{3}}{2}.\cos\theta - \frac{1}{2}.\sin\theta\right]$$

$$= 2\sqrt{2}\left[\cos\frac{\pi}{6}\cos\theta - \sin\frac{\pi}{6}\sin\theta\right]$$

$$= 2\sqrt{2}\cos\left[\theta + \frac{\pi}{6}\right]$$

$$\sqrt{6}.\cos\theta - \sqrt{2}\sin\theta = 2\sqrt{2}\cos\left[\theta + \frac{\pi}{6}\right]$$

$$(\sqrt{6}\cos\theta - \sqrt{2}\sin\theta)^2 = 8\cos^2\left[\theta + \frac{\pi}{6}\right] = 4$$

$$2\cos^2\left[\theta + \frac{\pi}{6}\right] = 1$$

$$2\cos^2\left[\theta + \frac{\pi}{6}\right] - 1 = 0$$

$$\cos 2\left[\theta + \frac{\pi}{6}\right] = 0$$

hence $2\left[\theta + \frac{\pi}{6}\right] = \frac{\pi}{2}$ or $\frac{3\pi}{2}$ or $\frac{5\pi}{2}$ ..... etc

$$2\theta = \frac{\pi}{2} - \frac{\pi}{3} \text{ or } \frac{3\pi}{2} - \frac{\pi}{3} \text{ or } \frac{5\pi}{2} - \frac{\pi}{3} \text{ .....}$$

$$\theta = \frac{\pi}{12} \text{ or } \frac{7\pi}{12} \text{ or } \frac{13\pi}{12} \ \ldots..$$

$$\text{in general} \quad \theta = \frac{\pi}{12} + \frac{6n\pi}{12} = \frac{\pi}{12} + \frac{n\pi}{2}$$

Question 3

Use identities for cos(C + D) and cos(C – D) to prove that

$$\cos A + \cos B = 2\cos\left[\frac{A+B}{2}\right].\cos\left[\frac{A-B}{2}\right]$$

Hence find in terms of π, the general solution of the equation

$$\cos 5\theta + \cos\theta = \cos 3\theta.$$

Using both the identity for cos A + cos B and the corresponding identity for sin A – sin B, show that

$$\sin 5\alpha - \sin\alpha = 2\sin\alpha(\cos 4\alpha + \cos 2\alpha).$$

The triangle PQR has angle QPR = α (which is not zero), angle PQR = 5α and RP = 3 RQ. Show that sin 5α = 3 sin α and deduce that

$$\cos 4\alpha + \cos 2\alpha = 1.$$

By solving a quadratic equation in cos 2α, or otherwise, find the value of α, giving your answer to the nearest one tenth of a degree.

AEB June 1986

Suggested solution

$$\cos(C + D) = \cos C.\cos D - \sin C.\sin D$$

$$\cos(C - D) = \cos C.\cos D + \sin C.\sin D$$

Let $A = (C + D)$

and $B = (C - D)$

hence $A + B = 2C$ and $A - B = 2D$

$C = \left[\frac{A+B}{2}\right]$ and $D = \left[\frac{A-B}{2}\right]$

thus $\cos A + \cos B = 2\cos C.\cos D$

$$= 2\cos\left[\frac{A+B}{2}\right].\cos\left[\frac{A-B}{2}\right]$$

$\cos 5\theta + \cos\theta = \cos 3\theta$

Let $A = 5\theta$ and $B = \theta$

$$\cos 5\theta + \cos\theta = 2\cos\left[\frac{5\theta+\theta}{2}\right].\cos\left[\frac{5\theta-\theta}{2}\right]$$

$$= 2\cos 3\theta.\cos 2\theta$$

hence $\quad 2\cos 3\theta.\cos 2\theta = \cos 3\theta$

At this stage resist the temptation to divide both sides of the equation by cos 3θ as solutions will be lost.

Instead factorise by cos 3θ.

$$\cos 3\theta[2\cos 2\theta - 1] = 0$$

solutions:

either $\quad \cos 3\theta = 0$ and $3\theta = \frac{\pi}{2}, \frac{3\pi}{2}, \frac{5\pi}{2}$ ..... etc

$$\theta = \frac{\pi}{6}, \frac{3\pi}{6}, \frac{5\pi}{6} \text{ ..... etc}$$

general solution $\theta = (2n + 1)\frac{\pi}{6}$

or $\quad 2\cos 2\theta - 1 = 0$ and $\cos 2\theta = \frac{1}{2}$

$$2\theta = \frac{\pi}{6}, \pi - \frac{\pi}{6}, \text{ ..... etc}$$

general solution $\quad \theta = n\pi \pm \frac{\pi}{6}$

$$\cos A + \cos B = 2\cos\left[\frac{A + B}{2}\right].\cos\left[\frac{A - B}{2}\right]$$

and $\quad \sin A - \sin B = 2\cos\left[\frac{A + B}{2}\right].\sin\left[\frac{A - B}{2}\right]$

so $\quad \sin 5\alpha - \sin \alpha = 2\cos 3\alpha . \sin 2\alpha$

$$= 2\cos 3a . 2\sin\alpha.\cos\alpha$$

$$= 2\sin\alpha\{2\cos 3\alpha.\cos\alpha\}$$

$$= 2\sin\alpha\{\cos 4\alpha + \cos 2\alpha\}$$

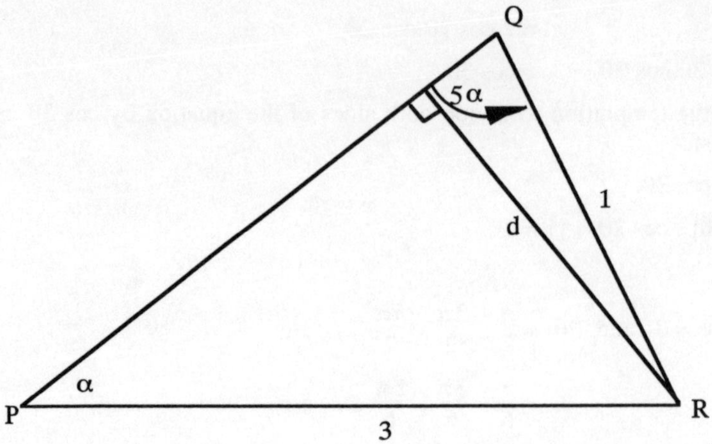

$\sin 5\alpha = \frac{d}{1}$ and $\sin \alpha = \frac{d}{3}$

$\therefore \quad d = \sin 5\alpha = 3 \sin \alpha$

Now $\quad \sin 5\alpha - \sin \alpha \quad = 2 \sin \alpha \, [\cos 4\alpha + \cos 2\alpha]$

hence $\quad 3 \sin \alpha - \sin \alpha \quad = 2 \sin \alpha \, [\cos 4\alpha + \cos 2\alpha]$

and $\quad 1 = \cos 4\alpha + \cos 2\alpha$

Expand $\cos 4\alpha = \cos [2\alpha + 2\alpha] \quad = 2 \cos^2 2\alpha - 1$

ie $\quad 2 \cos^2 2\alpha - 1 + \cos 2\alpha = 1$

$$2 \cos^2 2\alpha + \cos 2\alpha - 2 = 0$$

$$\cos 2\alpha = \frac{-(1) \pm \sqrt{(1)^2 - 4(2)(-2)}}{2(2)}$$

either $\quad \cos 2\alpha = \ 0.7807$ or $\cos 2\alpha = -1.280$

$2\alpha = 38.67^o$ impossible

$\alpha = 19.3^o$

Question 4

Given that $\sin^{-1}x$, $\cos^{-1}x$ and $\sin^{-1}(1-x)$ are acute angles,

i) prove that $\sin[\sin^{-1}x - \cos^{-1}x] = 2x^2 - 1$

ii) solve the equation $\sin^{-1}x - \cos^{-1}x = \sin^{-1}(1-x)$.

AEB June 1982

Suggested solution

i) Let $y = \sin^{-1}x$ and $z = \cos^{-1}x$

$x = \sin y \qquad x = \cos z$

hence
$$\begin{aligned}\sin(\sin^{-1}x - \cos^{-1}x) &= \sin(y - z)\\ &= \sin y.\cos z - \cos y.\sin z\\ &= (x)(x) - (\sqrt{1-x^2})(\sqrt{1-x^2})\\ &= x^2 - (1 - x^2)\\ &= 2x^2 - 1\end{aligned}$$

ii)
$$\sin^{-1}x - \cos^{-1}x = \sin^{-1}(1-x)$$

hence
$$\sin[\sin^{-1}x - \cos^{-1}x] = \sin[\sin^{-1}(1-x)]$$

$$2x^2 - 1 = \sin[\sin^{-1}(1-x)] = (1-x)$$

$$2x^2 + x - 2 = 0$$

$$x = \frac{-(-1) \pm \sqrt{(1)^2 - 4(2)(-2)}}{2(2)}$$

either
$$x = \frac{-1 + \sqrt{17}}{4} = 0.781$$

or
$$x = \frac{-1 + \sqrt{17}}{4} = 2.562$$

4 DIFFERENTIAL CALCULUS

4.1 Key formulae and ideas

1. Definition of a derivative

If $y = f(x)$ then $\dfrac{dy}{dx} = \lim\limits_{\partial x \to 0} \dfrac{f(x + \partial x) - f(x)}{\partial x}$

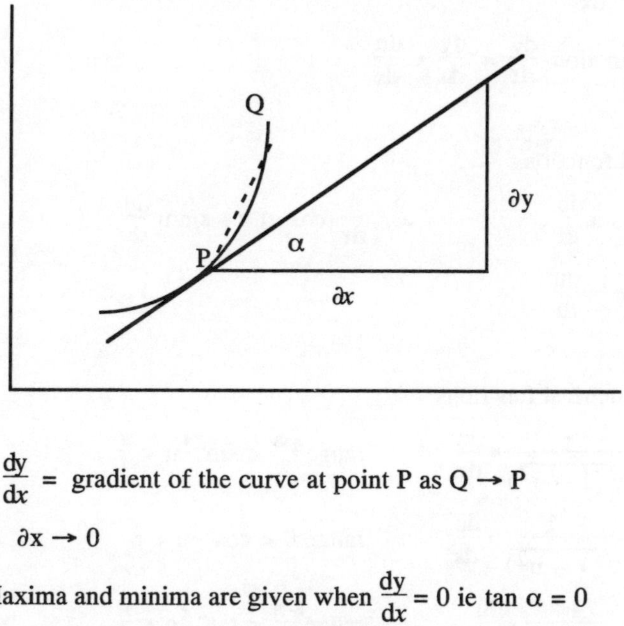

$\dfrac{dy}{dx}$ = gradient of the curve at point P as Q → P

ie $\partial x \to 0$

Maxima and minima are given when $\dfrac{dy}{dx} = 0$ ie $\tan \alpha = 0$

ie the tangent line to the curve is parallel to the x-axis.

2. Let u, v, w be functions of x; a, b, c, n are constants.

$$\frac{d}{dx}(a) = 0\,; \qquad \frac{d}{dx}(ax) = a\,; \qquad \frac{d}{dx}(ax^n) = nax^{n-1}$$

$$\frac{d}{dx}(u \pm v \pm w \pm .....) = \frac{du}{dx} \pm \frac{dv}{dx} \pm \frac{dw}{dx} \pm .....)$$

$$\frac{d}{dx}(au) = a\frac{du}{dx}$$

Product $\quad \frac{d}{dx}(uv) = u\frac{dv}{dx} + v\frac{du}{dx}$

Quotient $\quad \frac{d}{dx}\left[\frac{u}{v}\right] = \frac{v(du/dx) - u(dv/dx)}{v^2}$

$$\frac{d}{dx}(u^n) = n.u^{n-1}\frac{du}{dx}$$

Function of a function $\quad \frac{dy}{dx} = \frac{dy}{du} \bullet \frac{du}{dx}$

3. Trigonometrical functions

$$\frac{d}{dx}(\sin u) = \cos u.\frac{du}{dx}\,; \qquad \frac{d}{dx}(\cos u) = -\sin u.\frac{du}{dx}$$

$$\frac{d}{dx}(\tan u) = \sec^2 u.\frac{du}{dx}$$

4. Inverse trigonometrical functions

$$\frac{d}{dx}(\sin^{-1}u) = \frac{1}{\sqrt{(1-u^2)}} \bullet \frac{du}{dx} \qquad \text{range } \frac{-\pi}{2} < \sin^{-1}u < \frac{\pi}{2}$$

$$\frac{d}{dx}(\cos^{-1}u) = \frac{-1}{\sqrt{(1-u^2)}} \bullet \frac{du}{dx} \qquad \text{range } 0 < \cos^{-1}u < \pi$$

$$\frac{d}{dx}(\tan^{-1}u) = \frac{1}{(1+u^2)} \bullet \frac{du}{dx} \qquad \text{range } \frac{-\pi}{2} < \tan^{-1}u < \frac{\pi}{2}$$

5. Exponential and logarithmic functions

$$\frac{d}{dx}(\log_e u) = \frac{1}{u} \bullet \frac{du}{dx}$$

$$\frac{d}{dx}(\log_a u) = \frac{d}{dx}\left[\frac{\log_e u}{\log_e a}\right] = \frac{1}{\log_e a} \bullet \frac{1}{u} \bullet \frac{du}{dx} \quad \{a \neq 0, 1\}$$

$$\frac{d}{dx}(e^u) = e^u \bullet \frac{du}{dx}$$

6. Higher derivatives

$$\frac{d}{dx}\left[\frac{dy}{dx}\right] = \frac{d^2y}{dx^2} \qquad \{ = f''(x) = y'' \}$$

7. To determine maxima and minima

i) from $y = f(x)$

ii) find $\frac{dy}{dx}$

iii) put $\frac{dy}{dx} = 0$ ..... solve for x ... at this value of x there is a turning point.

iv) to determine whether a maximum or minimum

find $\frac{d^2y}{dx^2}$ ... substitute value(s) for x

if a negative value ..... maximum

if a positive value ..... minimum

v) substitute for x in $y = f(x)$ and determine the y coordinate(s).

4.2 Questions and solutions

Question 1

Given that $x = e^{3t} \cos 3t$ and $y = e^{3t} \sin 3t$

show that $\dfrac{dy}{dx} = \tan\left(3t + \dfrac{\pi}{4}\right)$

Cambridge November 1986

Suggested solution

Differentiate both y and x with respect to t using the product function.

$$\frac{dy}{dt} = e^{3t}\,\{3\}\cos(3t) + \sin(3t)\,\{3\}\,e^{3t}$$

$$= 3\,e^{3t}\,\{\cos(3t) + \sin(3t)\}$$

$$\frac{dx}{dt} = e^{3t}\,(-3)\sin(3t) + \cos(3t).(3)\,e^{3t}$$

$$= 3\,e^{3t}\,\{\cos(3t) - \sin(3t)\}$$

$$\frac{dy}{dx} = \frac{dy}{dt} \bullet \frac{dt}{dx} = \frac{\cos(3t) + \sin(3t)}{\cos(3t) - \sin(3t)}$$ divide top and bottom by cos(3t)

$$= \frac{1 + \tan(3t)}{1 - \tan(3t)}$$ bear in mind that $\tan\left(\dfrac{\pi}{4}\right) = 1$

$$= \frac{\tan(\pi/4) + \tan(3t)}{1 - \tan(3t).\tan(\pi/4)}$$

$$= \tan\left(3t + \frac{\pi}{4}\right)$$ cf expansion tan(A + B) page 31

Question 2

a) Differentiate with respect to x

i) $\log_{10}x$

ii) $\sec(x^2)$

b) Given that y = A cos 2x + B sin 4x, where A and B are constants, prove that

$$\frac{d^4y}{dx^4} + 20\frac{d^2y}{dx^2} + 64y = 0$$

AEB June 1984

Suggested solution

a)

i) Let $y = \log_{10}x = \frac{1}{\log_e(10)} \bullet \log_e x$ cf page 2

$$\frac{dy}{dx} = \frac{1}{\log_e(10)} \bullet \frac{1}{x} = \frac{1}{x.\log_e(10)}$$

ii) Let $y = \sec(x^2)$ put $z = x^2$

$$\frac{dz}{dx} = 2x$$

hence $y = \sec(z) = \{\cos(z)\}^{-1}$

$$\frac{dy}{dz} = (-1)\cos^{-2}(z).\{-\sin(z)\} \qquad = +\frac{\sin(z)}{\cos^2 z}$$

$$= \sec(z).\tan(z)$$

thus $\frac{dy}{dx} = \frac{dy}{dz} \bullet \frac{dz}{dx} = 2x.\sec(x^2).\tan(x^2)$

b)

$$y = A.\cos(2x) + B.\sin(4x)$$

$$\frac{dy}{dx} = -2A\sin(2x) + 4B\cos(4x)$$

$$\frac{d^2y}{dx^2} = -4A\cos(2x) - 16B\sin(4x)$$

$$\frac{d^3y}{dx^3} = 8A\sin(2x) - 64B\cos(4x)$$

$$\frac{d^4y}{dx^4} = 16A\cos(2x) + 256B\sin(4x)$$

hence $$\frac{d^4y}{dx^4} + 20\frac{d^2y}{dx^2} + 64y$$

$=>$ $$16A\cos(2x) + 256B\sin(4x) - 80A\cos(2x) - 320B\sin(4x)$$
$$+ 64A\cos(2x) + 64\sin(4x) = 0$$

Question 3

Find the values of x for which the function $f(x) = 4x^3 + 6x^2 - 9x + 2$ has a maximum or minimum. Draw a rough sketch of the graph of the function $f(x)$, indicating clearly the positions of any maximum or minimum.

Oxford June 1983

Suggested solution

$$f(x) = 4x^3 + 6x^2 - 9x + 2$$

$$f'(x) = 12x^2 + 12x - 9$$

To locate any maximum or minimum put

$$f'(x) = 0 = 12x^2 + 12x - 9$$

factorise $(2x + 3)(2x - 1) = 0$

either $2x + 3 = 0$ or $2x - 1 = 0$

$x = -\frac{3}{2}$ $\qquad$ $x = \frac{1}{2}$

to determine whether maximum or minimum find f"(x)

$$f''(x) = 24x + 12$$

when $x = -\frac{3}{2}$ $f''(x) = -36 + 12 = -24$ ..... maximum

when $x = \frac{1}{2}$ $f''(x) = +12 + 12 = +24$ ..... minimum

values of y

put $x = -\frac{3}{2}$ $\qquad f(x) = 4\left[\frac{-3}{2}\right]^3 + 6\left[\frac{-3}{2}\right]^2 - 9\left[\frac{-3}{2}\right] + 2 = 15\frac{1}{2}$

put $x = \frac{1}{2}$ $\qquad f(x) = 4\left[\frac{1}{2}\right]^3 + 6\left[\frac{1}{2}\right]^2 - 9\left[\frac{1}{2}\right] + 2 = -\frac{1}{2}$

Other values of y:	$x = -2$,		0 ,	+1
	$y = +12$,	+2,	+3	

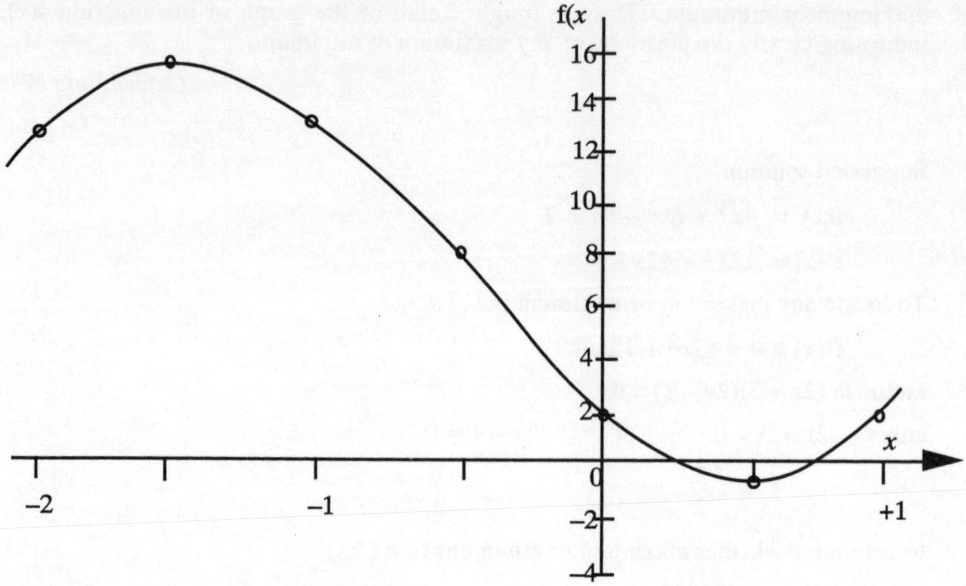

Question 4

For a given mass of gas, the volume, V cm^3, and the pressure, p cm of mercury, are related by $p = k V^n$ where k and n are constants.

a) Prove that $\frac{dp}{dV} = \frac{np}{V}$.

b) For a particular mass of gas n = – 1.4. At the instant when the volume is 20 cm^3 the pressure is 150 cm of mercury and the volume is decreasing at the rate of 0.5 cm^3s^{-1}. Calculate the rate of change of pressure at this instant in cm of mercury per second.

c) The following measurements of the volume V cm^3 and the pressure, p cm of mercury, of a given mass of gas were taken

V	10	50	110	170	230
p	1412.5	151.4	50.3	27.4	18.6

By plotting values of $\log_{10}p$ against V, verify graphically the relationship

$p = k v^n$ where k and n are constants.

Use your graph to find approximate values for k and n, giving your answers to two significant figures.

AEB June 1987

Suggested solution

a) $p = k.V^n$ and $\frac{dp}{dV} = n.k.V^{n-1}$

but $V^{n-1} = \frac{p}{k.V}$ $\therefore$ $\frac{dp}{dV} = \frac{n.p}{V}$

b) Using function of a function differentiation

Rate of change of pressure $= \frac{dp}{dt} = \frac{dp}{dV} \cdot \frac{dV}{dt}$

when $n = -1.4$,

$p = 150$ cm Mercury, $V = 20\ cm^3$, $\frac{dV}{dt} = -0.5$ cm2s^{-1}

$$\frac{dp}{dt} = \frac{(-1.4)(150)(-0.5)}{20} = +5.25 \text{ cm Mercury } s^{-1}$$

c) $p = k.V^n$

$$\log_{10}p = n.\log_{10}V + \log_{10}k$$

Compare this equation with the standard form for a straight line graph

$y = m.x + c$

ie $\log_{10}p$ corresponds to y and $\log_{10}V$ to x.

Gradient of the line is given by n and the intercept on the $\log_{10}p$ vertical axis by $\log_{10}k$.

V	$\log_{10}V$	p	$\log_{10}p$
10	1.000	1412.5	3.1500
50	1.6990	151.4	2.1801
110	2.0414	50.3	1.7016
170	2.2304	27.4	1.4378
230	2.3617	18.6	1.2695

i) $\log_{10} k = 4.45$

$\therefore \quad k = 28183.8$

ie $\quad k = 28000$ to 2 sig fig

ii) n = gradient

$$= -\frac{(4.45 - 0)}{(3.35 - 0)}$$

$= -1.33$

$n = -1.3$ to 2 sig fig

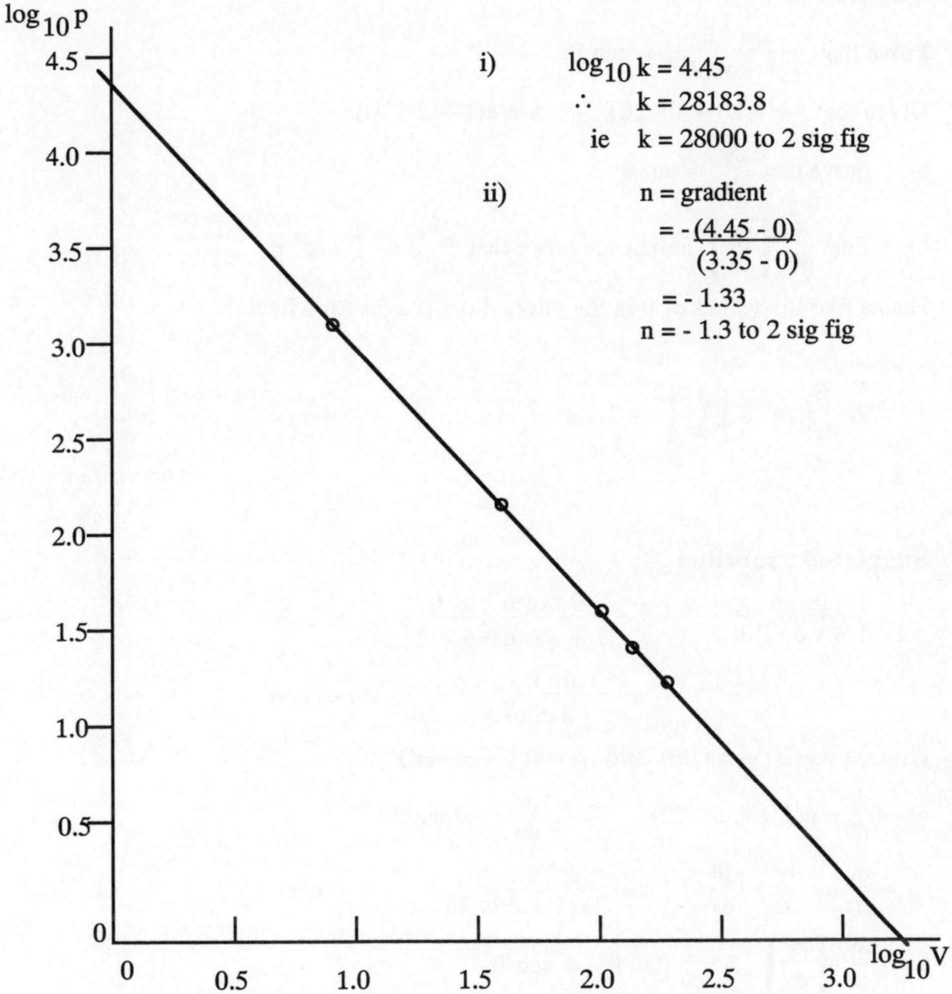

Question 5

Prove that $\dfrac{\sin 2\theta}{1 + \cos 2\theta} = \tan\theta$.

Given that $x = a(2\theta + \sin 2\theta)$, $\quad y = a(1 - \cos 2\theta)$,

a) prove that $\dfrac{dy}{dx} = \tan\theta$

b) find $\dfrac{d}{d\theta}\left[\dfrac{dy}{dx}\right]$ and hence prove that $\dfrac{d^2y}{dx^2} = \dfrac{1}{4a}\sec^4\theta$.

Hence find the values of θ in the interval $0 \le \theta \le 2\pi$ for which

$$2a\frac{d^2y}{dx^2} = 3\left[\frac{dy}{dx}\right]^2 - 1.$$

Oxford June 1987

Suggested solution

$$\frac{\sin 2\theta}{1 + \cos 2\theta} = \frac{2\sin\theta.\cos\theta}{1 + \{2\cos^2\theta - 1\}}$$

$$= \frac{2\sin\theta.\cos\theta}{2\cos^2\theta} \qquad = \tan\theta$$

Given $x = a(2\theta + \sin 2\theta)$ and $y = a(1 - \cos 2\theta)$

a) $\dfrac{dx}{d\theta} = a(2 + 2.\cos 2\theta) \qquad \dfrac{dy}{d\theta} = a.2.\sin 2\theta$

$$\frac{dy}{dx} = \frac{dy}{d\theta} \cdot \frac{d\theta}{dx} = \frac{a\{2\sin 2\theta\}}{2a\{1 + \cos 2\theta\}} = \tan\theta$$

b) $\dfrac{d}{d\theta}\left[\dfrac{dy}{dx}\right] = \dfrac{d}{d\theta}\{\tan\theta\} = \sec^2\theta$

$$\frac{d^2y}{dx^2} = \frac{d}{d\theta}\left[\frac{dy}{dx}\right] \cdot \frac{d\theta}{dx} = \frac{\sec^2\theta}{2a\{1 + \cos 2\theta\}}$$

$$= \frac{1}{2a} \cdot \sec^2\theta \cdot \frac{\sin 2\theta}{2\{1 + \cos 2\theta\}\sin 2\theta}$$

$$= \frac{1}{4a} \cdot \frac{1}{\cos^2\theta} \cdot \frac{\tan\theta}{\sin\theta.\cos\theta}$$

$$= \frac{1}{4a} \cdot \sec^4\theta$$

$$2a\frac{d^2y}{dx^2} = 3\left[\frac{dy}{dx}\right]^2 - 1$$

$$\frac{2a}{4a\cos^4\theta} = 3(\tan\theta)^2 - 1$$

$$\frac{\sec^4\theta}{2} = \frac{\{\tan^2\theta + 1\}^2}{2} = 3\tan^2\theta - 1$$

$$\tan^4\theta - 4\tan^2\theta + 3 = 0$$

$$(\tan^2\theta - 3)(\tan^2\theta - 1) = 0$$

either $\tan^2\theta - 3 = 0$ or $\tan^2\theta - 1 = 0$

$\tan^2\theta = 3$ $\qquad$ $\tan^2\theta = 1$

$\tan\theta = \pm\sqrt{3}$ $\qquad$ $\tan\theta = \pm 1$

take + sign $\theta = 60^o,\ 240^o$ $\qquad$ $\theta = 45^o,\ 225^o$

take – sign $\theta = 120^o,\ 300^o$ $\qquad$ $\theta = 135^o,\ 315^o$

θ has values $\frac{\pi}{4}, \frac{\pi}{3}, \frac{2\pi}{3}, \frac{3\pi}{4}, \frac{5\pi}{4}, \frac{4\pi}{3}, \frac{5\pi}{3}, \frac{7\pi}{4}$

5 DIFFERENTIAL CALCULUS

5.1 Key formulae and ideas

u, v, w are functions of x ; a, b, c are constants ;

e is the natural base of logarithms = 2.7183

$\log_e u$ or $l_n u$ is the natural logarithm of u ;

all angles are expressed in radians;

arbitrary constants of integration are omitted.

Standard integrals

$\int a \, dx = x$

$\int a.f(x) \, dx = a \int f(x) \, dx$

$\int (u \pm v \pm w \pm ...) \, dx = \int u \, dx \pm \int v \, dx \pm \int w \, dx \pm ...$

$\int u \frac{dv}{dx} = uv - \int v \frac{du}{dx}$ integration by parts

$\int u^n \, du = \frac{u^{n+1}}{n + 1}$ where n is not equal to –1

$\int \frac{du}{u} = \log_e u$ if u > 0

$= -\log_e u$ if u < 0

$= \log_e | u |$

$\int e^u \, du = e^u$

$\int a^u \, du = \int e^{ulna} \, du = \frac{e^{ul_n a}}{l_n a} = \frac{a^u}{lna}$ a > 0, a ≠ 1

$\int \sin u \, du = \cos u$

$\int \cos u \, du = - \sin u$

$\int \sec^2 u \, du = \tan u$

$$\int \frac{dx}{x^2 + a^2} = \frac{1}{a} \tan^{-1} \left[\frac{x}{a}\right]$$

$$\int \frac{dx}{\sqrt{(a^2 - x^2)}} = \sin^{-1} \left[\frac{x}{a}\right]$$

$$\int \frac{dx}{(x^2 - a^2)} = \frac{1}{2a} \log_e \left[\frac{x - a}{x + a}\right] \quad \text{if } x^2 > a^2$$

$$= \frac{1}{2a} \log_e \left[\frac{a + x}{a - x}\right] \quad \text{if } x^2 < a^2$$

Methods of integration

1. $\int \sin^2 x \, dx = \int \frac{1}{2}(1 - \cos 2x) = \frac{1}{2}\left(x - \frac{1}{2}\sin 2x\right)$

2. Likewise $\int \cos^2 x \, dx$

3. $\int \log_e x \, dx$; this requires special treatment ie

   consider $\frac{d}{dx}(x \log_e x) = (1) \log_e(x) + x \left[\frac{1}{x}\right]$

   $= \log_e (x) + 1$

   hence $x \log_e x = \int (\log_e x + 1) \, dx$

   $\therefore$ $\int \log_e x \, dx = x \log_e x - \int dx$

   $\int \log_e x \, dx = x \log_e x - x$

4. Quotients where denominator contains a function of x

   a) where degree of x in numerator > degree of x in denominator ..... divide and then integrate.

   b) if numerator can be made equal to derivative of denominator ..... integral is the $\log_e$ of the denominator.

   c) denominator can be factorised ..... partial fractions.

5. By substitution or change of variable

   eg if $y = \int f(u) \, du$ and $\frac{dy}{du} = f(u)$

then $\quad \dfrac{dy}{dx} = \dfrac{dy}{du} \cdot \dfrac{du}{dx} = f(u)\,\dfrac{du}{dx}$

thus $\quad y = \int f(u)\,\dfrac{du}{dx} \cdot dx$

6. $\int \sin^a x \cos x \, dx$ ..... note that $\cos x$ is the differential of $\sin x$

$\rightarrow \dfrac{1}{a+1} \sin^{a+1} x$

7. $\int \sin^m x \cos^n x \, dx$ can ALWAYS be treated by substitution if

EITHER m or n (or both) is an ODD, POSITIVE INTEGER

ie m is odd ..... let $u = \cos x$

n is odd ...... let $u = \sin x$

8. Any rational function of $\sin x$, $\cos x$ etc can be transformed into a rational algebraic function by substituting

$u = \tan \left[\dfrac{x}{2}\right]$

9. An irrational function of the form $\sqrt{(ax + b)}$ can be rationalised by substituting

$u^2 = ax + b$

10. Approximate integration ..... giving area under curve where $y = f(x)$

Trapezium rule: $\quad \text{area} = \dfrac{1}{2} \{y_1 + 2y_2 + 2y_3 \ldots + 2y_{n-1} + y_n\}\, d$

Simpson's rule: $\quad \text{area} = \dfrac{1}{3}\, d\, \{y_1 + 4y_2 + y_3\}$

5.2 Questions and solutions

Question 1

Evaluate

a) $\int_0^{\frac{\pi}{3}} \sin^2 x \cos x \, dx$

b) $\int_1^2 \frac{5 - x}{(x + 1)(2x - 1)} dx$

Oxford June 1987

Suggested solution

a) Note that cos x dx is the differential of sin x

$$\int_0^{\frac{\pi}{3}} \sin^2 x \, . \cos x \, dx = \left[\frac{1}{3}\sin^3 x\right]_0^{\frac{\pi}{3}}$$

$$= \frac{1}{3}\left[[\sin\frac{\pi}{3}]^3 - \sin(0)^3\right]$$

$$= \frac{1}{3}\left[\left(\frac{\sqrt{3}}{3}\right)^3 - 0\right]$$

$$= \frac{\sqrt{3}}{8}$$

b) The denominator factorises ..... use partial fractions

$$\frac{5 - x}{(x + 1)(2x - 1)} = \frac{A}{(x + 1)} + \frac{b}{(2x - 1)}$$

multiply both sides of the equation by the denominator

$$5 - x = A(2x - 1) + B(x + 1)$$

try x = –1 ; $\quad 5 - (-1) = A[2(-1) - 1] = 0$

ie $\quad A = 3$

ie B = 3

$$\text{thus} \int_1^2 \frac{(5-x)}{(x+1)(2x-1)}\,dx = \int_1^2 \frac{3}{(x+1)}\,dx + \int_1^2 \frac{3}{(2x-1)}\,dx$$

Note that the numerator is the differential of the denominator

$$= 3[\log_e(x+1)]_1^2 + \frac{3}{2}\int_1^2 \frac{2}{(2x-1)}\,dx$$

$$= 3[\log_e(x+1)]_1^2 + \frac{3}{2}[\log_e(2x-1)]_1^2$$

$$\text{ie} \int_1^2 \frac{(5-x)}{(x+1)(2x-1)}\,dx$$

$$= \frac{3}{2}[[2\log_e(2+1) - 2\log e\,(1+1)] + 3[\log e(4-1) - \log_e(2-1)]]$$

$$= \frac{3}{2}[5\log_e(3) - 2\log_e(2) - 3\log_e(1)]$$

$$= \frac{3}{2}\left[\log_e\left[\frac{(3)^5}{(2)^2}\right]\right] = \frac{3}{2}\log_e\left[\frac{243}{4}\right] = 6.16$$

Question 2

a) Find $\int \sin 2x.(\sin x)^{\frac{1}{2}}\,dx$.

b) Prove that $\displaystyle\int_1^2 \frac{dx}{x^3 + x} = \frac{1}{2}\log_e\left(\frac{8}{5}\right)$.

AEB June 1984

Suggested solution

a) $\displaystyle\int \sin 2x.(\sin x)^{\frac{1}{2}}\,dx = \int 2.\sin x \,.\, \cos x \,.\, (\sin x)^{\frac{1}{2}}\,dx$

$$= \int 2 \sin^{\frac{3}{2}}(x) \,.\, \cos(x)\,dx$$

$$= \frac{4}{5}\sin^{\frac{5}{2}}(x)$$

b) $\displaystyle\int_1^2 \frac{dx}{x^3 + x} = \int_1^2 \frac{dx}{x(x^2 + 1)}$

use partial fractions

$$\frac{1}{x(x^2 + 1)} = \frac{A}{x} + \frac{Bx + C}{x^2 + 1}$$

ie $\quad 1 = A(x^2 + 1) + (Bx + C)x$

put $x = 0$: $\quad 1 = A$

put $x = 1$: $\quad 1 = (1)(2) + (B[1] + C)(1)$ ..... $\quad B + C = -1$

put $x = 2$: $\quad 1 = (1)(5) + (2B + C)(2)$ ..... $\quad 4B + 2C = -4$

$\quad 2B + C = -2$

hence $\quad (2B + C) - (B + C) = -2 - (-1)$

$\quad B = -1$ and $C = 0$

$$\int_1^2 \frac{dx}{x(x^2+1)} = \int_1^2 \left[\frac{1}{x} - \frac{x}{(x^2+1)}\right] dx$$

$$= \left[\log_e x - \frac{1}{2}\log_e(x^2+1)\right]_1^2$$

$$= \left[\log_e 2 - \frac{1}{2}\log_e 5\right] - \left[\log_e 1 - \frac{1}{2}\log_e 2\right]$$

$$= \frac{1}{2}\left[\log_e(2)^3 - \log_e 5\right]$$

$$= \frac{1}{2}\log_e\left[\frac{8}{5}\right]$$

Question 3

Evaluate the integrals

a) $\displaystyle\int_0^1 \frac{1}{x+1}\,dx$

b) $\displaystyle\int_0^1 \frac{x}{x+1}\,dx$

c) $\displaystyle\int_0^1 \frac{1}{\sqrt{2x+1}}\,dx$

Oxford June 1986

Suggested solution

a) $\displaystyle\int_0^1 \frac{1}{x+1}\,dx$ the numerator is the differential of the denominator

$$= \left[\log_e (x+1)\right]_0^1$$

$$= \log_e (1+1) - \log_e(0+1) = \log_e 2 = 0.6931$$

b) $\displaystyle\int_0^1 \frac{x}{x+1}\,dx$ divide the numerator by the denominator

$$= \int_0^1 \left[1 - \frac{1}{x+1}\right] dx$$

$$= \left[x - \log_e(x+1)\right]_0^1$$

$$= [1 - \log_e 2] - [0 - \log_e 1]$$

$$= 1 - \log_e 2 = 0.3069$$

c) $$\int_0^1 \frac{1}{\sqrt{(2x + 1)}}\,dx = \int_0^1 [2x + 1]^{-\frac{1}{2}}\,dx$$

$$= \left[\frac{2}{2}(2x + 1)^{\frac{1}{2}}\right]_0^1$$

$$= \left[2\left(\frac{1}{2}\right) + 1\right] - [2(0) + 1]$$

$$= 2 - 1 = 1$$

Question 4

Evaluate the integral $\displaystyle\int_0^3 \sqrt{25 - x^2}\, dx$

The region A is bounded by the line y = 4 and the smaller arc of the circle $x^2 + y^2 = 25$. Use your previous result to find the area of A.

A solid sphere of radius 5 units has a cylindrical hole of radius 4 units bored through it, such that the axis of the cylinder passes through the centre of the sphere.

Determine the volume of the part of the sphere which remains, leaving your answer in terms of π.

Oxford June 1986

Suggested solution

$\int \sqrt{25 - x^2}\, dx$ — use a trigonometrical substitution

let $x = 5 \sin \theta$

$\frac{dx}{d\theta} = 5 \cos \theta$

$= \int \sqrt{(25 - 25 \sin^2\theta)}\,.\,5 \cos \theta\,.d\theta$

$= \int 5\sqrt{(1 - \sin^2\theta)}\,.\,5 \cos \theta\,.\,d\theta$ — note $\sin^2\theta + \cos^2\theta = 1$

$= 25 \int \cos \theta\,.\,\cos \theta\,.\,d\theta$ — note $\cos 2\theta = 2 \cos^2\theta - 1$

$= \frac{25}{2} \int (1 + \cos 2\theta)\, d\theta$

$= \frac{25}{2}\left(\theta + \frac{1}{2} \sin 2\theta\right)$ — expand $\sin 2\theta$

$= \frac{25}{2} (\theta + \sin \theta\,.\,\cos \theta)$ — revert from θ to x

$= \frac{25}{2}\left[\sin^{-1} \frac{x}{5} + \frac{x}{5}\sqrt{(1 - \frac{x^2}{a^2})}\right]$ insert limits $x = 3$, $x = 0$

$= \frac{25}{2}\left[\sin^{-1} \frac{3}{5} + \frac{3}{5}\sqrt{(1 - \frac{9}{25})} - 0 - 0\right]$

$= \frac{25}{2}$ [0.6435 + 0.48] remember the angle is in radians

= 14.04

Region A

Circle $x^2 + y^2 = 25$ has centre (0, 0) and radius 5.

In the diagram the required area A is shaded and is the area under the arc from $x = 0$ to $x = 3$ less the area of the rectangle labelled B.

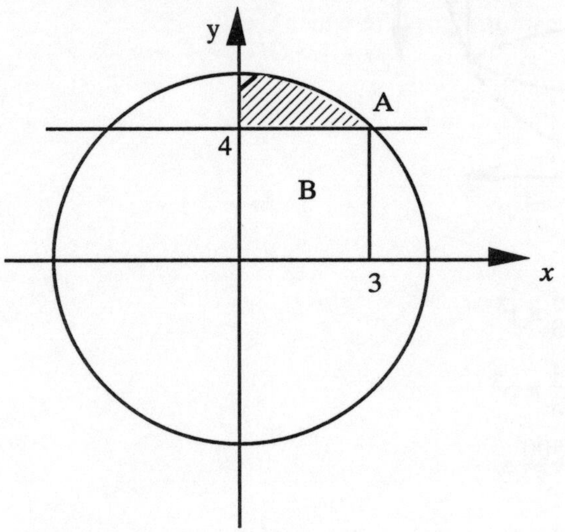

Area under arc $= \int_0^3 y.dx$

$= \int_0^3 \sqrt{(25 - x^2)}\, dx$

Required area = 14.04 – (4 × 3)

= 2.04

Sphere and cylinder

As seen from the diagram the height of the cylinder is $2 \times 3 = 6$ units.

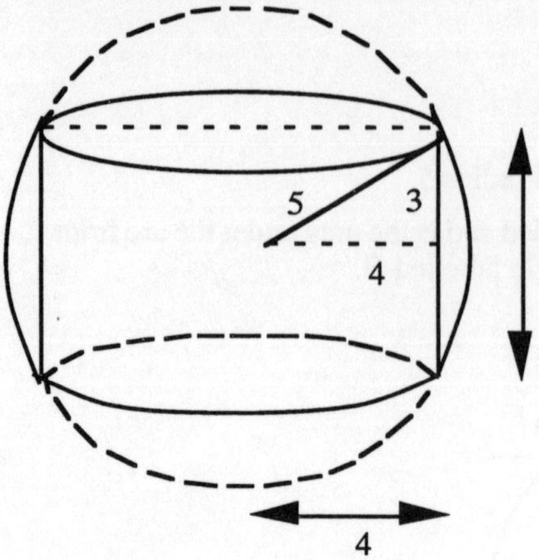

Volume of sphere $= \frac{4}{3}\pi r^3$

$= \frac{4}{3}\pi 5^3$

$= \frac{500}{3}\pi$

Volume of cylinder $= \pi r^2 h$

$= \pi 4^2 . 6$

$= 96\pi$

Volume of remainder $= \frac{500}{3}\pi - 96\pi$

$= 212\pi$

Question 5

Showing your working in the form of a table, use Simpson's rule with 4 equal intervals to estimate.

$$\int_0^2 \ln (1 + x^2)\, dx ,$$

giving your answer to 3 decimal places.

Deduce an approximate value for

$$\int_0^2 \ln \sqrt{(1 + x^2)}\, dx .$$

AEB June 1987

Suggested solution

Between the limits 0 and 2, four equal intervals require ordinates at the x values 0, 0.5, 1.0, 1.5 and 2.0 as shown in the sketch curve.

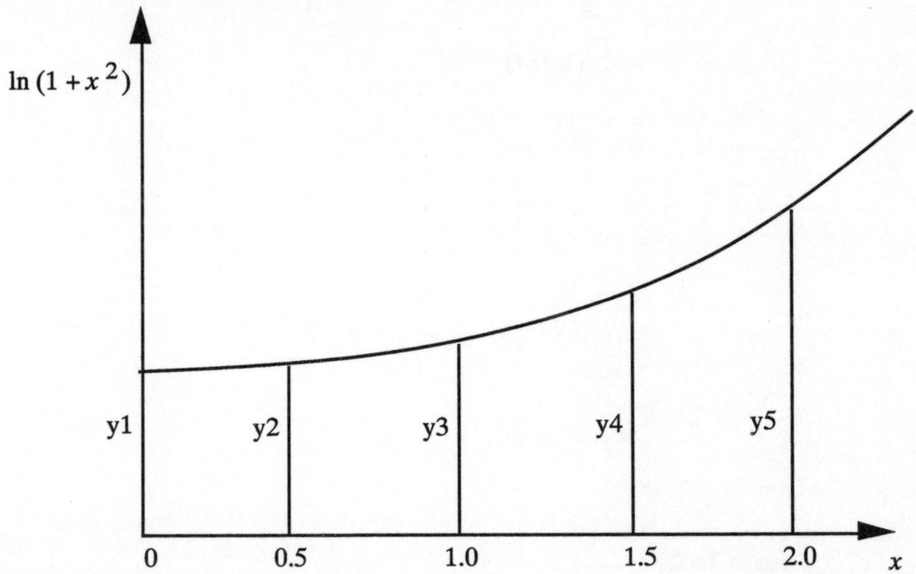

x	0	0.5	1.0	1.5	2.0
$1 + x^2$	1	1.25	2.0	3.25	5.0
$\ln(1 + x^2)$	0	0.2231	0.6931	1.1787	1.6094

Apply Simpson's rule twice:

$$\text{Area} = \frac{1}{3}(d)[\,y_1 + 4y_2 + y_3\,] + \frac{1}{3}(d)[\,y_3 + 4y_4 + y_5\,]$$

$$\text{Area} = \frac{1}{3}(0.5)[0 + 4(0.2231) + 2(0.6931) + 4(1.1787) + 1.6094]$$

$$= 1.4338 = 1.434 \text{ to 3 dec places.}$$

$$\int_0^2 \ln(\sqrt{1 + x^2}) = \int_0^2 \ln(1 + x^2)^{\frac{1}{2}} = \int_0^2 \frac{1}{2}\ln(1 + x^2)$$

$$= \frac{1}{2}(1.434)$$

$$= 0.717$$

Question 6

A solid right circular cone of greatest volume is to be cut from a solid sphere of radius 3R. Using calculus, show that the height of this cone is 4R and calculate its volume in terms of R.

AEB June 1983

Suggested solution

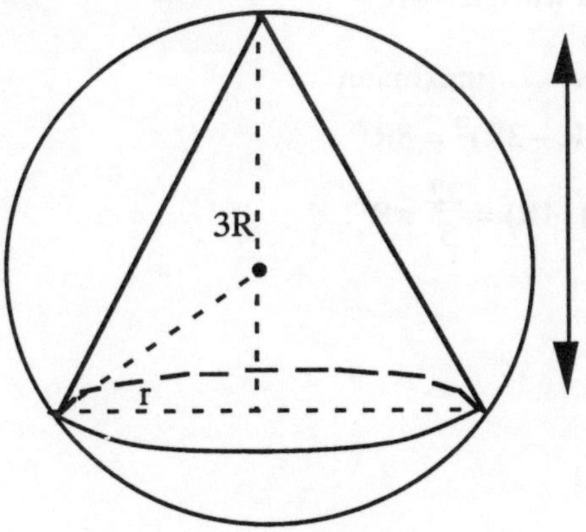

Volume of cone $= \frac{1}{3}\pi r^2 h$

from the diagram, radius of base of cone is given by

$$r^2 = (3R)^2 - (h - 3R)^2$$

substituting for r

$$\text{volume} = V = \frac{1}{3}\pi h\,(9R^2 - [h^2 - 6Rh + 9R^2])$$

$$= 2\pi Rh^2 - \frac{1}{3}\pi h^3$$

differentiate V with respect to h

$$\frac{dV}{dh} = 4\pi Rh - \pi h^2$$

to determine a maximum put $\frac{dV}{dh} = 0$

ie $h(4\pi R - \pi h) = 0$

either $h = 0$ not a sensible solution

or $h = 4R$

to show that $h = 4R$ represents a maximum obtain the second differential.

ie $\frac{d^2V}{dh^2} = 4\pi R - 2\pi h$ and when $h = 4R$

$= -4\pi R$ ..... negative ..... maximum

Radius of base of cone $r^2 = (3R)^2 - (4R - 3R)^2 = 8R^2$

Maximum volume of cone $= \frac{1}{3}\pi(8R^2)(4R) = \frac{32}{3}\pi R^2$.

6 COORDINATE GEOMETRY

6.1 Key formulae and ideas

1. Standard equation of straight line.

$y = m.x + c$

where: m = gradient or slope = $\tan \theta$

c = intercept on y–axis

$-\left[\frac{c}{m}\right]$ = intercept on x–axis

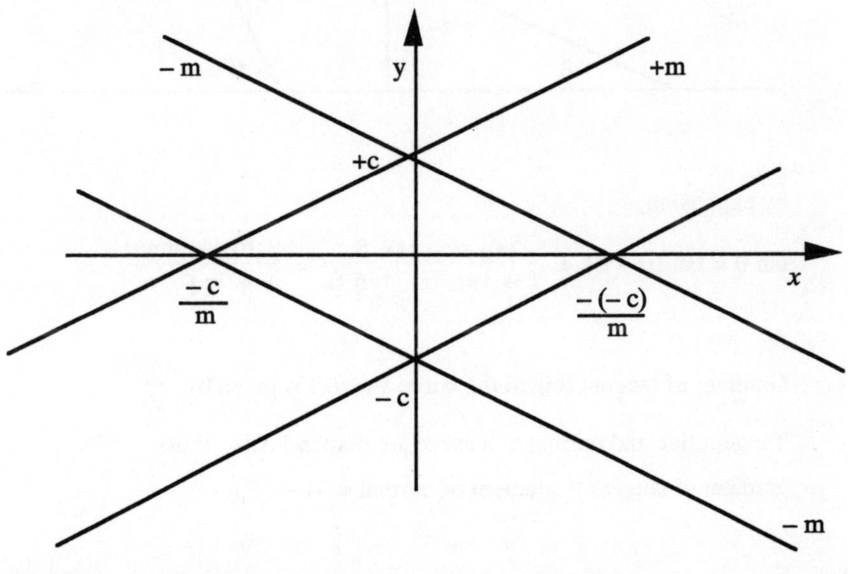

Note: any equation which can be reduced to this linear pattern ie both x and y variables to the power of 1 will give a straight line graph.

2. When two straight lines are perpendicular to each other, the product of the gradients = –1.

   ie $m_1 \cdot m_2 = -1$

   when two straight lines are parallel to each other

   $$m_1 = m_2$$

3. Angle between two straight lines

   $\theta = \alpha - \beta$

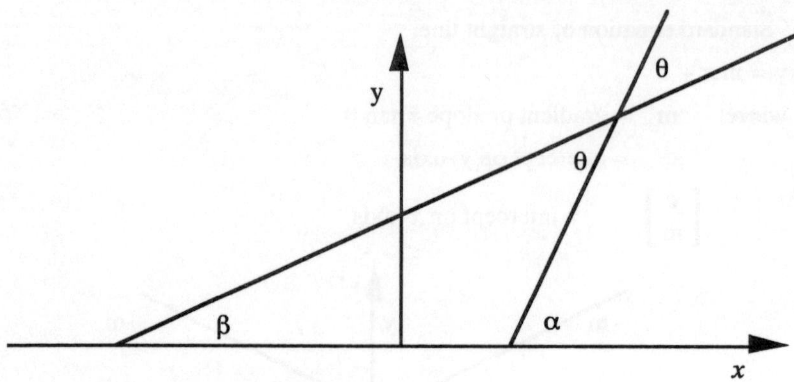

   by expansion

   $$\tan\theta = \tan(\alpha - \beta) = \frac{\tan\alpha - \tan\beta}{1 + \tan\alpha \cdot \tan\beta} = \frac{m_1 - m_2}{1 + m_1 \cdot m_2}$$

4. Gradient of tangent line to the curve $y = f(x)$ is given by $\frac{df(x)}{dx}$

   Tangent line and normal to a curve are perpendicular, thus

   gradient of tangent × gradient of normal = –1

5. Definition of a Circle .. locus of a point which maintains a constant distance from a fixed point.

   Standard equation of a Circle

   $x^2 + y^2 + 2gx + 2fy + c = 0$

   has centre at coordinates (–g, –f)

   and radius = $\sqrt{g^2 + f^2 - c}$

6. Definition of a Parabola .. the locus of a point P which is equidistant from a fixed point S and a fixed line L.

   Standard equation of a Parabola

   $y^2 = 4ax$

   S (focus), coordinates (a, 0)

   L (directrix)

   PM = PS

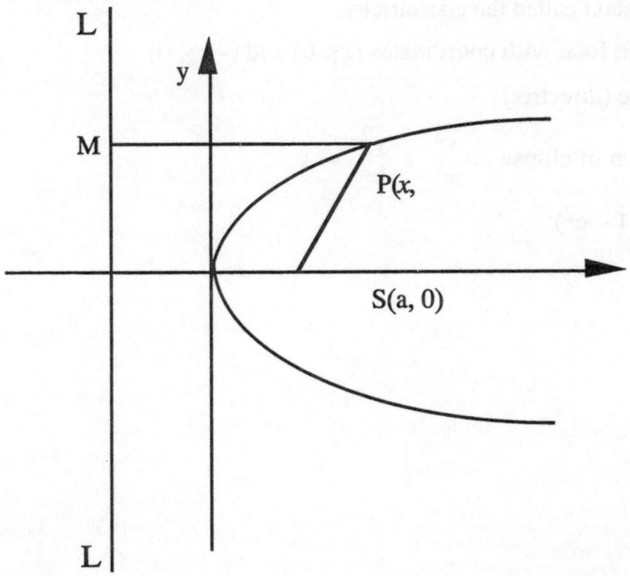

7. Ellipse ... the locus of a point where the ratio of its distance from a fixed point to its distance from a fixed line is constant and less than 1

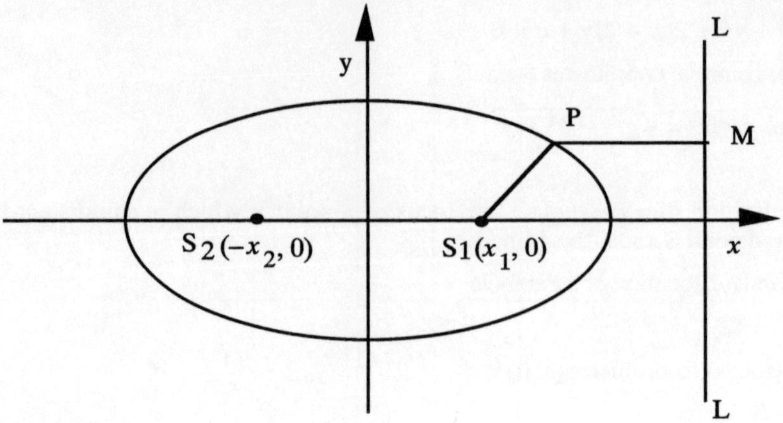

ie $\frac{PS}{PM} = e < 1$

where e is a constant called the eccentricity

S_1 and S_2 are the focii with coordinates $(x_1, 0)$ and $(-x_2, 0)$

L is the fixed line (directrix)

Standard equation of ellipse ... $\frac{x^2}{a^2} + \frac{y^2}{b^2} = 1$

where $b^2 = a^2(1 - e^2)$

8. Definition of a Hyperbola ... locus of a point where the ratio of its distance from a fixed point to its distance from a fixed line is a constant and greater than 1.

$\frac{PS}{PM} = e > 1$

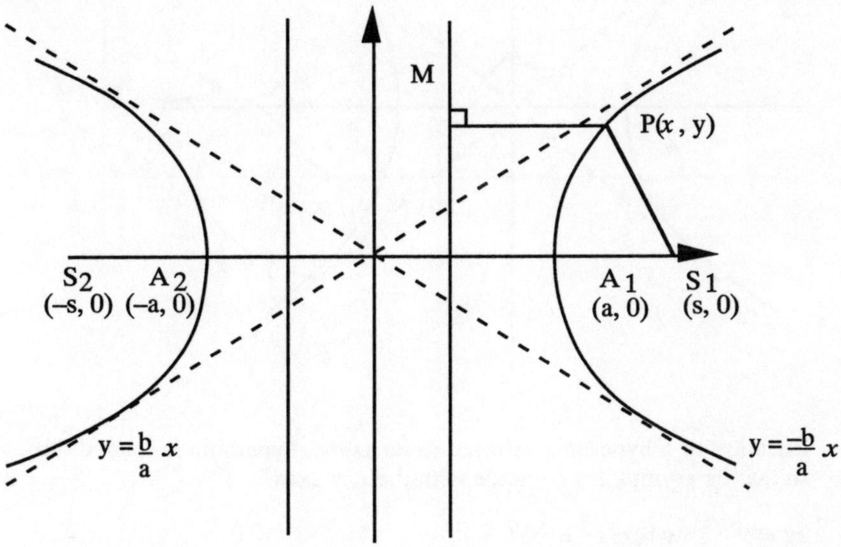

Standard equation of a Hyperbola

$$\frac{x^2}{a^2} - \frac{y^2}{b^2} = 1 \qquad \text{where } b^2 = a^2(e^2 - 1)$$

Asymptotes $\qquad y = \pm \frac{b}{a} . x$

Rectangular Hyperbola ... the asymptotes are at right angles to each other

ie $b = a$

equation

$$x^2 - y^2 = a^2$$

$$c^2 = \frac{1}{2} a^2$$

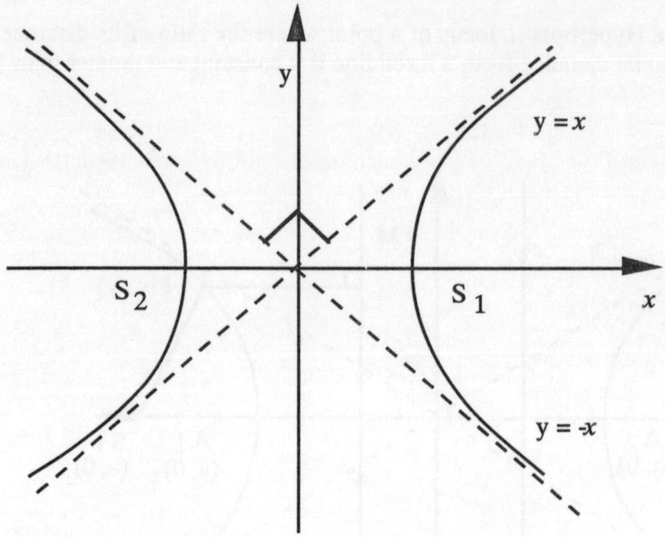

Equation of a hyperbola referred to its axes... hyperbola is rotated through 45° so that its asymptotes coincide with the x, y axes.

$xy = c^2$ where $c^2 = \frac{1}{2} a^2$

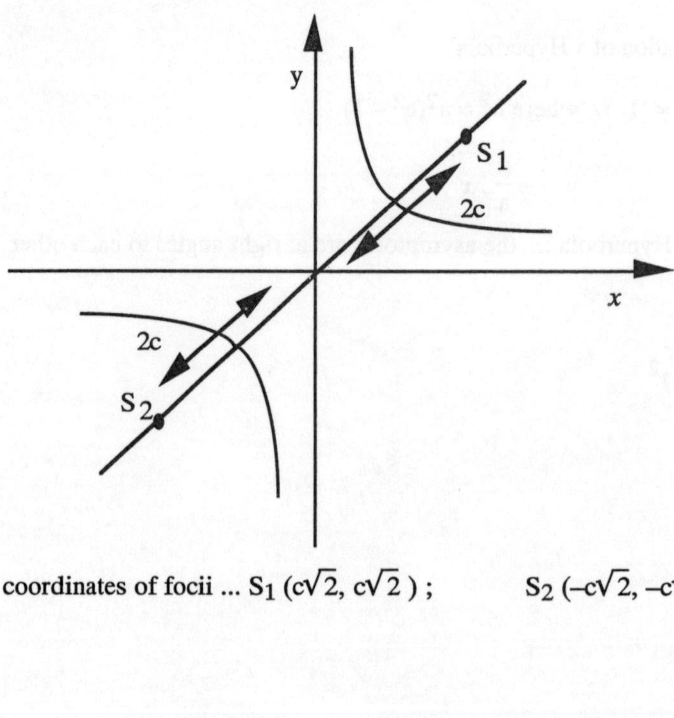

coordinates of focii ... $S_1\,(c\sqrt{2}, c\sqrt{2})$; $S_2\,(-c\sqrt{2}, -c\sqrt{2})$

6.2 Questions and solutions

Question 1

Two circles are drawn in the plane $0xy$ and each passes through the origin 0. The first circle has centre (a,0) and the second circle has centre (0,b).

The line $y = x \tan \theta$ cuts the first circle again at P and the second circle again at Q. Show that M, the mid-point of PQ, is at

$\{\cos \theta\, [a \cos \theta + b \sin \theta], \sin \theta\, [a \cos \theta + b \sin \theta]\}$.

Verify that M lies on the circle

$$x^2 + y^2 - ax - by = 0$$

AEB June 1984

Suggested solution

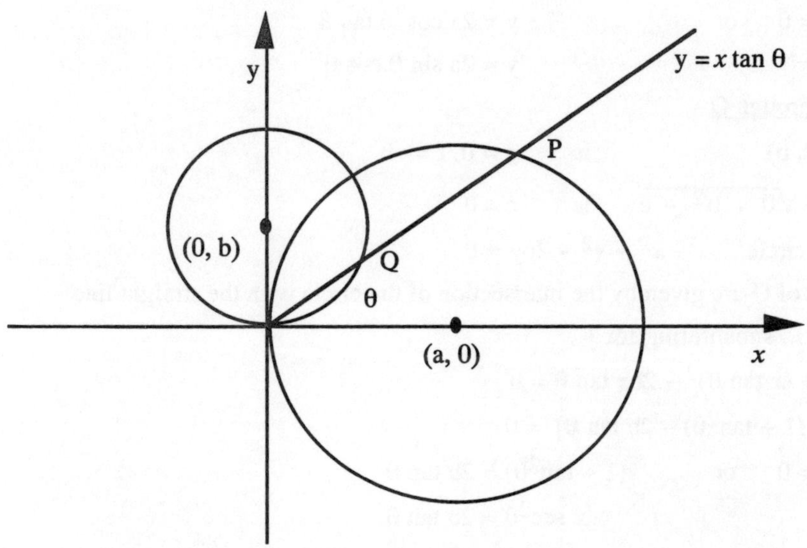

General equation of a circle

$$x^2 + y^2 + 2gx + 2fy + c = 0$$

centre $\Rightarrow (-g, -f)$;   radius $\Rightarrow \sqrt{g^2 + f^2 - c}$

1st circle through P

centre ⇒ (a, 0) ie $g = -a$ and $f = 0$

radius ⇒ $a = \sqrt{(-a)^2 + 0 - c}$ ie $c = 0$

equation of circle: $x^2 + y^2 - 2ax = 0$

coordinates of P are given by the intersection of the circle with the straight line $y = x \tan \theta$ ...

Substituting for y

$$x^2 + (x \tan \theta)^2 - 2ax = 0$$

$$x[x(1 + \tan^2\theta) - 2a] = 0$$

either $x = 0$ or $x(1 + \tan^2\theta) = 2a$

$$x \sec^2\theta = 2a$$

$$x = 2a \cos^2\theta$$

substituting in $y = x \tan \theta$

either $y = 0$ or $y = 2a \cos^2\theta.\tan \theta$

$$y = 2a \sin \theta.\cos \theta$$

2nd circle through Q

centre ⇒ (0, b) ie $g = 0, f = -b$

radius $= b = \sqrt{0 + b^2 - c}$ ie $c = 0$

equation of circle: $x^2 + y^2 - 2by = 0$

coordinates of Q are given by the intersection of the circle with the straight line $y = x \tan \theta$ ... substituting for y

$$x^2 + (x \tan \theta)^2 - 2bx \tan \theta = 0$$

$$x[x(1 + \tan^2\theta) - 2b \tan \theta] = 0$$

either $x = 0$ or $x(1 + \tan^2\theta) = 2b \tan \theta$

$$x \sec^2\theta = 2b \tan \theta$$

$$x = 2b \sin \theta.\cos \theta$$

substituting for x in $y = x \tan \theta$

either $y = 0$ or $y = (2b \sin^2\theta).\tan \theta$

$$= 2b \sin^2\theta$$

<u>Mid-point of PQ = M</u>

x coordinate is given by $\frac{P_x + Q_x}{2}$

$$= \frac{2a \cos^2\theta + 2b \sin\theta.\cos\theta}{2}$$

ie $$x = \cos\theta\,(a.\cos\theta + b\sin\theta)$$

y coordinate is given by $\frac{Py + Qy}{2}$

$$= \frac{2a \sin\theta.\cos\theta + 2b \sin^2\theta}{2}$$

ie $$y = \sin\theta\,(a.\cos\theta + b.\sin\theta)$$

<u>Circle through M</u> $\quad x^2 + y^2 - ax - by = 0$

but M lies on line $y = x\tan\theta$ hence substituting

$$x^2 + (x\tan\theta)^2 - ax - b(x\tan\theta) = 0$$

factorise $\quad x[x + x\tan^2\theta - a - b\tan\theta] = 0$

either $\quad x = 0$ ... not possible

or $\quad x + x\tan^2\theta - a - b\tan\theta = 0$

substitute x coordinate for the point M

$$\cos\theta\,(a\cos\theta + b\sin\theta)\,(1 + \tan^2\theta) - (a + b\tan\theta) = 0$$

$$\cos\theta\,(a\cos\theta + b\sin\theta).\sec^2\theta - (a + b\tan\theta) = 0$$

divide the first term by $\cos^2\theta$

$$(a + b\tan\theta) - (a + b\tan\theta) = 0$$

ie left hand side = right hand side = 0 which is satisfied if point M lies on the circle

$$x^2 + y^2 - ax - by = 0$$

Question 2

Prove that the circle with equation

$$x^2 + y^2 - 2ax - 2by + a^2 = 0$$

touches the x-axis.

Find the equations of the two circles which pass through the points (2,1) and (3,2) and touch the x-axis.

AEB June 1983

Suggested solution

General equation of a circle

$$x^2 + y^2 + 2gx + 2fy + c = 0$$

centre $\Rightarrow (-g, -f)$; radius $\Rightarrow \sqrt{g^2 + f^2 - c}$

for the given circle

$$2g = -2a \quad \text{and} \quad 2f = -2b \quad \text{and} \quad c = a^2$$

ie $g = -a$ $f = -b$

coordinates of centre $\Rightarrow \{-(-a), -(-b)\} \Rightarrow \{a, b\}$

radius $\Rightarrow \sqrt{(-a)^2 + (-b)^2 - a^2} \Rightarrow b$

hence circle touches the x-axis.

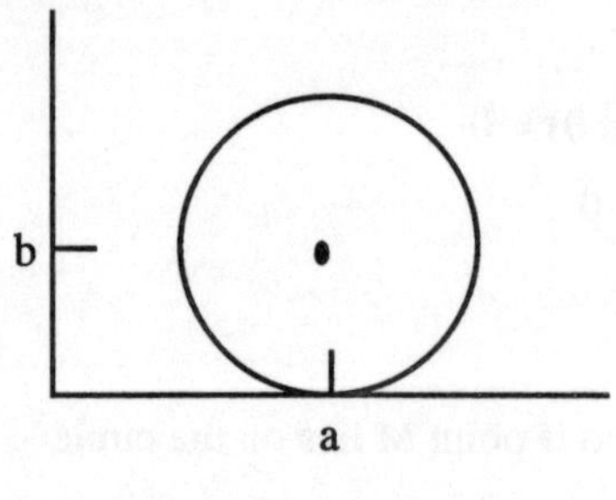

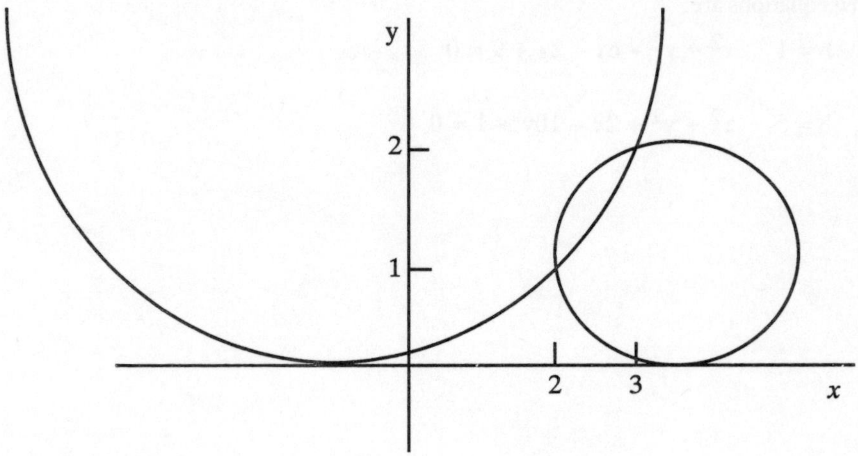

The required circles each touch the x-axis and therefore have the equation $x^2 + y^2 - 2ax - 2by + a^2 = 0$. There will be two values for each of a and b. These will be obtained by setting up a pair of simultaneous equations for each of the points (2, 1) and (3, 2).

at (2, 1): $2^2 + 1^2 - 2a(2) - 2b(1) + a^2 = 0$

ie $5 - 4a - 2b + a^2 = 0$ ..... (i)

at (3, 2): $3^2 + 2^2 - 2a(3) - 2b(2) + a^2 = 0$

ie $13 - 6a - 4b + a^2 = 0$ ..... (ii)

to solve multiply equation (i) by 2

$10 - 8a - 4b + 2a^2 = 0$ ..... (iii)

subtract equation (iii) from (ii)

$3 + 2a - a^2 = 0$

ie $a^2 - 2a - 3 = 0$

$(a - 3)(a + 1) = 0$

either $a - 3 = 0$ or $a + 1 = 0$

$a = 3$ $a = -1$

in equation (i)

$5 - 4(3) - 2b + 3^2 = 0$ or $5 - 4(-1) - 2b + (-1)^2 = 0$

$b = 1$ $b = 5$

Required equations are:

a = 3, b = 1 $x^2 + y^2 - 6x - 2y + 9 = 0$

]

a = –1, b = 5 $x^2 + y^2 + 2x - 10y + 1 = 0$

Question 3

Prove that the equation of the normal to the parabola

$y^2 = 4ax$ at the point P $(at^2, 2at)$ is

$$y + tx = 2at + at^3.$$

The normal at P meets the x-axis at G and the line through P parallel to the y-axis meets the x-axis at H. Prove that the length of GH is constant as t varies.

The mid-point of PG is M. The line through M perpendicular to PG cuts PH at Q. Prove that the coordinates of Q are $(at^2, at - a/t)$ and determine a cartesian equation for the locus of Q as t varies.

Oxford June 1987

Suggested solution

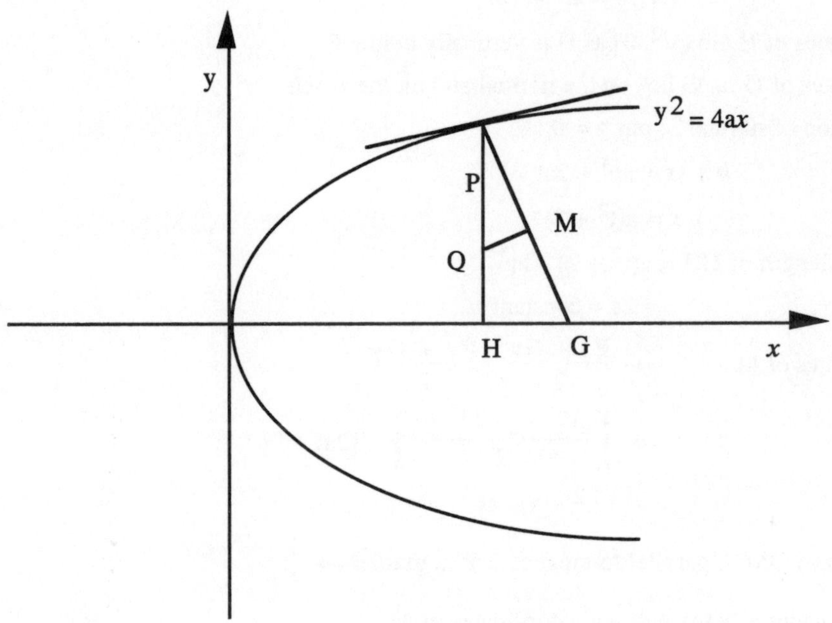

Equation of parabola: $y^2 = 4ax$

To obtain gradient of tangent line ... differentiate with respect to x

$$2y \frac{dy}{dx} = 4a \qquad \text{ie} \qquad \frac{dy}{dx} = \frac{2a}{y}$$

At point P $(at^2, 2at)$ $\quad \frac{dy}{dx} = \frac{2a}{2at} = \frac{1}{t}$

gradient of normal × gradient of tangent $= -1$

gradient of normal $= -t$

equation of normal .... $y = mx + c$

at point P $(at^2, 2at)$

$$2at = (-t).at^2 + c$$

$$c = at^3 + 2at$$

ie equation of normal $\quad y = -tx + at^3 + 2at$

$$y + tx = at^3 + 2at$$

Coordinates of H are $(at^2, 0)$ as H is vertically below P

coordinates of G .... G lies on the normal and on the x-axis

in equation of normal ... put $y = 0$

$$0 + tx = at^3 + 2at$$

$$x = a(t^2 + 2)$$

hence length of GH $= a(t^2 + 2) - at^2$

$$= 2a = \text{constant}$$

Coordinates of M $\quad \Rightarrow \frac{Px + Gx}{2}, \frac{Py + Gy}{2}$

$$\Rightarrow \left[\frac{at^2 + a(t^2 + 2)}{2}\right], [2at + 0]$$

$$\Rightarrow (at^2 + a), at$$

Equation of QM ... parallel to tangent at P ... gradient $= \frac{1}{t}$

ie using $y = mx + c$ and coordinates of M

$$at = \frac{1}{t}(at^2 + a) + c \text{ and } c = \frac{-a}{t}$$

ie

$$y = \frac{1}{t}.x - \frac{a}{t}$$

at Q ... x coordinate = at^2 as Q is vertically below P

and $\quad y = \frac{1}{t}(at^2) - \frac{a}{t} = at - \frac{a}{t}$

Locus of Q ... eliminate t from the x and y coordinates and obtain an equation in x and y.

$$x = at^2 \quad \ldots\ldots \quad t = \sqrt{\frac{x}{a}}$$

$$y = at - \frac{a}{t} = \frac{at^2 - a}{t} = (x - a)\frac{\sqrt{a}}{\sqrt{x}}$$

square $\quad y^2 = (x - a)^2 \cdot \frac{a}{x}$

Question 4

Prove that the tangent at P(4, 4) to the curve $4y = x^2$ has equation

$$2x - y - 4 = 0$$

This tangent meets the line $4x + 3y - 12 = 0$ at the point Q.

Calculate the coordinates of Q.

The normal at P to the curve $4y = x^2$ meets the line $4x + 3y - 12 = 0$ at the point R.

Calculate the coordinates of the point R.

Show that the circle on QR as diameter has equation

$$x^2 + (y - 4)^2 = 16$$

Sketch in the same diagram the curve $4y = x^2$ and the circle $x^2 + (y - 4)^2 = 16$ and write on your sketch the coordinates of any points where the two curves meet or intersect.

AEB June 1987

Suggested solution

Tangent at P ... to obtain the gradient differentiate equation $4y = x^2$ with respect to x.

$$4\frac{dy}{dx} = 2x \quad ..... \quad \frac{dy}{dx} = \frac{x}{2}$$

at point P(4, 4) $\qquad \frac{dy}{dx} = 2$

equation of tangent $\qquad y = mx + c$

$$4 = 2(4) + c \quad ..... \quad c = -4$$

ie $\qquad y = 2x - 4 \quad ..... \quad \text{(i)}$

Coordinates of Q .... given by intersection of this tangent with line

$$4x + 3y - 12 = 0 \; ..... \quad \text{(ii)}$$

multiply equation (i) by 2 $\qquad 4x - 2y - 8 = 0$

subtract $\qquad 5y - 4 = 0$

$$y = +\frac{4}{5}$$

Substitute for y in (i) ... $\qquad \frac{4}{5} = 2x - 4 \text{ ie } \quad x = +\frac{12}{5}$

Normal at P

gradient of normal × gradient of tangent $= -1$

gradient of normal $= -\frac{1}{2}$

equation of normal ... $y = mx + c$

$4 = -\frac{1}{2}(4) + c$ ..... $c = +6$

ie $y = -\frac{1}{2}x + 6$

$x + 2y - 12 = 0$ ..... (iii)

Coordinates of R ... given by the intersection of the normal (iii) with line equation (ii)

ie $4x + 3y - 12 = 0$

multiply equation (iii) by 4 ... $4x + 8y - 48 = 0$

subtract $5y - 36 = 0$

$y = \frac{36}{5}$

and in equation (iii)

$x + \frac{2(36)}{5} - 12 = 0$ ie $x = -\frac{12}{5}$

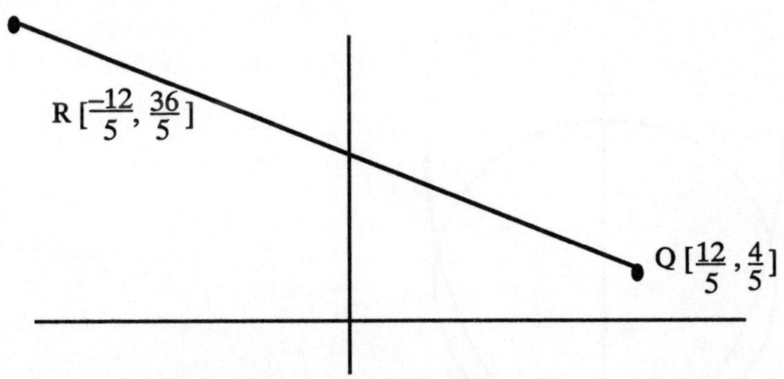

equation of circle ... $x^2 + y^2 + 2gx + 2fy + c = 0$

centre of circle is at the mid-point of QR

ie $$-g = \left[\frac{\frac{12}{5} + \frac{-12}{5}}{2}\right] = 0 \quad \text{and} \quad -f = \left[\frac{\frac{4}{5} + \frac{36}{5}}{2}\right] = 4$$

in equation of circle using coordinates of Q to find c

$$\left[\frac{12}{5}\right]^2 + \left[\frac{4}{5}\right]^2 + 0 + 2(-4)\left(\frac{4}{5}\right) + c = 0 \quad \text{ie } c = 0$$

hence equation of circle $\quad x^2 + y^2 - 8y = 0$

in form of equation $\quad x^2 + y^2 - 8y + 16 = 16$

$$x^2 + (y - 4)^2 = 16$$

Circle passes through point $\quad x = 0,\ y = 0$

Points of intersection of curve and circle

circle $\quad x^2 + y^2 - 8y = 0$

curve $\quad x^2 - 4y = 0$

subtract $\quad y^2 - 4y = 0 \quad \ldots\ldots \quad y(y - 4) = 0$

either $\quad y = 0 \quad$ or $\quad y = +4$

ie $\quad x^2 - 4(0) = 0 \qquad x^2 - 4(4) = 0$

$\quad x = 0 \qquad x = \pm 4$

Points of contact $\quad x = 0, \quad y = 0$

intersection $\quad x = -4, \quad y = +4$

$\quad x = +4, \quad y = +4$

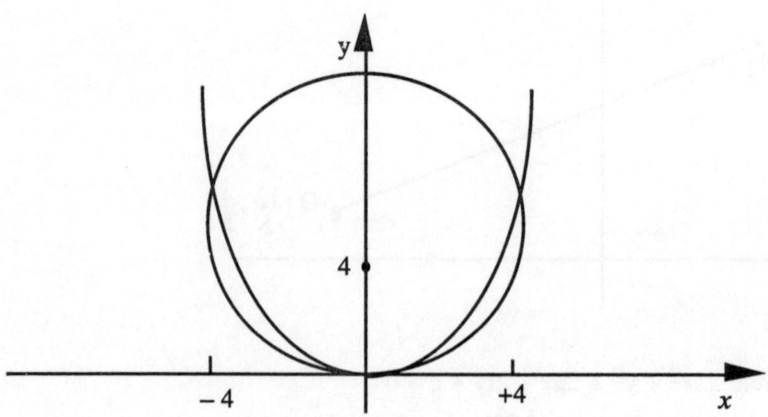

Question 5

The points P and Q lie on the hyperbola $xy = c^2$ and have coordinates (cp, c/p) and (cq, c/q) respectively. Obtain the equation of the chord PQ and show that the equation of the tangent to the hyperbola at the point P is $x^2 + p^2y = 2cp$.

The chord PQ passes through the origin: show that p + q = 0.

The tangent at P meets the axes 0x and 0y at the points A and B respectively. The lines QA and QB meet the hyperbola again at the points R and S respectively, where the coordinates of R and S are (cr, c/r) and (cs, c/s).

Find the values of r and s and show that RS is parallel to AB.

Oxford June 1986

Suggested solution

$xy = c^2$

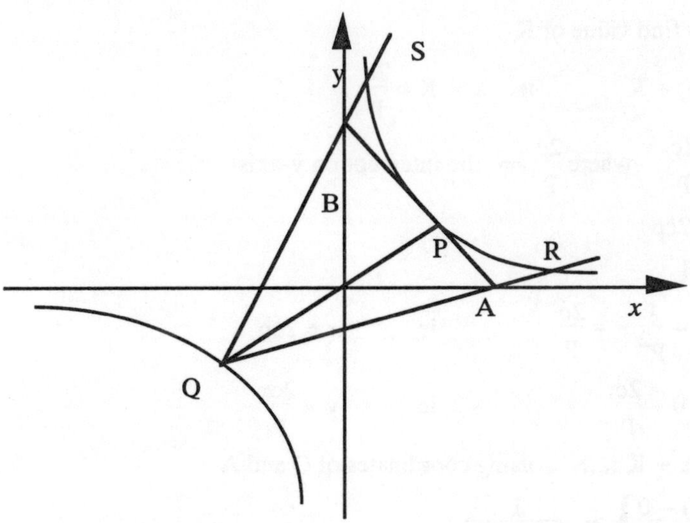

Equation of chord $y = mx + K$

$$\text{gradient} = m = \frac{(c/p - c/q)}{(cp - cq)} = \frac{(q - p)}{pq(p - q)} = -\frac{1}{pq}$$

value of K using coordinates of P

$$\frac{c}{p} = -\frac{1}{pq}(cp) + K \quad \text{ie} \quad K = \frac{c}{p} + \frac{c}{q} = \frac{c(p+q)}{pq}$$

equation of chord PQ $\quad y = -\frac{1}{pq}(x) + \frac{c(p+q)}{pq}$

$$pqy = -x + c(p+q)$$

Chord passes through origin

$$0 = 0 + c(p+q)$$

either $c = 0$ ..... not possible or $\quad p + q = 0$

equation of tangent ..... $y = mx + K$ ..... gradient is obtained by differentiating $xy = c^2$ with respect to x

$$x \cdot \frac{dy}{dx} + y = 0 \qquad \text{ie} \qquad \frac{dy}{dx} = -\frac{y}{x}$$

use coodinates of P to find value of K

$$\frac{c}{p} = \frac{-(c/p)}{(cp)} \cdot (cp) + K \qquad \text{ie} \qquad K = \frac{2c}{p}$$

hence $y = -\frac{1}{p^2} \cdot x + \frac{2c}{p}$ $\quad$ where $\frac{2c}{p}$ $\Rightarrow$ the intercept on y-axis

ie $\qquad p^2y = -x + 2cp$

coordinates of A and B

at A, $y = 0 \qquad 0 = -\frac{1}{p^2}x + \frac{2c}{p} \qquad$ ie $\qquad x = 2cp$

at B, $x = 0 \qquad y = 0 + \frac{2c}{p} \qquad$ ie $\qquad y = \frac{2cx}{p}$

line QAR ..... $y = mx + K$ ..... $\qquad$ using coordinates of Q and A

$$\text{gradient} = m = \left[\frac{(c/q) - 0}{cq - 2cp}\right] = \frac{1}{q(q-2p)}$$

value of K using coordinates of Q

$$\frac{c}{q} = \frac{1}{q(q-2p)} \cdot cp + K \qquad \text{ie} \qquad K = -\frac{2cp}{q(q-2p)}$$

hence line QAR $y = \dfrac{1}{q(q-2p)}x - \dfrac{2cp}{q(q-2p)}$

coordinates of R $\dfrac{c}{r} = \dfrac{1}{q(q-2p)}cr - \dfrac{2cp}{q(q-2p)}$

$$q(q-2p) = r^2 - 2pr$$

$r^2 - 2pr - q(q-2p) = 0$ ..... a quadratic equation

$$r = \frac{-(-2p) \pm \sqrt{(-2p)^2 - 4(1)(-q[q-2p])}}{2(1)}$$

$$= \frac{2p \pm \sqrt{(4p^2 + 4q^2 - 8pq)}}{2} = \frac{2p \pm 2\sqrt{(p-q)^2}}{2}$$

$$= p \pm (p-q)$$

either $r = p + (p-q)$ or $r = p - (p-q)$

$r = 2p - q$ $\quad$ $r = q$ ... position of Q

line QBS ... $y = mx + k$ ... using coordinates of Q and B

$$\text{gradient} = m = \left[\frac{(c/q) - (2c/p)}{cq - 0}\right] = \frac{(p-2q)}{pq^2}$$

value of k using coordinates of Q

$$\frac{c}{q} = \frac{(p-2q)}{pq^2}(cq) + k \qquad \text{ie} \qquad k = \frac{2c}{p}$$

hence line QBS $y = \dfrac{(p-2q)}{pq^2}x + \dfrac{2c}{p}$

coordinates of S $\dfrac{c}{s} = \dfrac{(p-2q)}{pq^2}(cs) + \dfrac{2c}{p}$

$$pq^2 = (p-2q)s^2 + 2sq^2$$

$(p-2q)s^2 + 2q^2s - pq^2 = 0$ ..... a quadratic equation

$$s = \frac{-2q^2 \pm \sqrt{(2q^2)^2 - 4(p-2q)(-pq^2)}}{2(p-2q)}$$

$$= \frac{-2q^2 \pm \sqrt{4q^4 + 4p^2q^2 - 8pq^3}}{2(p-2q)}$$

$$= \frac{-2q^2 \pm 2q\sqrt{p^2 - 2pq + q^2}}{2(p-2q)} = \frac{-q^2 \pm q(p-q)}{(p-2q)}$$

either $s = \dfrac{-q^2 + pq - q^2}{(p-2q)}$ or $s = \dfrac{-q^2 - pq + q^2}{(p-2q)}$

$= \dfrac{pq - 2q^2}{(p-2q)}$ $\qquad = -\dfrac{pq}{(p-2q)}$

$= \dfrac{q(p-2q)}{(p-2q)}$

$= q$ ... position of Q

gradient of AB $= \dfrac{0 - (2c/p)}{2cp - 0} = -\dfrac{2c}{2cp^2} = -\dfrac{1}{p^2}$

gradient of RS $= \left[\dfrac{(c/r) - (c/s)}{cr - cs}\right] = \dfrac{[s - r]}{rs[r - s]} = -\dfrac{1}{rs}$

$$= -\frac{1}{(2p-q)(-)\dfrac{pq}{(p-2q)}}$$

as $p + q = 0$

$p = -q \qquad = \dfrac{p - 2(-p)}{[2p - (-p)]\, p(-p)}$

$$= \frac{3p}{3p\,(-p^2)}$$

$$= -\frac{1}{p^2}$$

ie line RS is parallel to line AB having the same gradient.

7 APPROXIMATIONS AND SERIES

7.1 Key formulae and ideas

1. Newton-Raphson approximation for root of $f(x)$

   If α is a first rough approximation, then a second approximation is given by

   $$\alpha - \frac{f(\alpha)}{f'(\alpha)}$$

   The process can then be repeated to find other approximate values.

2. Proof by method of induction

   Show that if a summation of a series holds for a value of r, it holds for a value of (r + 1).

   Verify numerically by choosing a particular value for r, say r = 1.

3. Expansion of $\log_e (1 + z) = z - \frac{z^2}{2} + \frac{z^3}{3} \; ..... \pm \frac{z^n}{n}$

   + sign if n is an odd number

   – sign if n is an even number

   expansion is valid only if $-1 < z \leq +1$

7.2 Questions and solutions

Question 1

Find the coordinates of the turning points on the graph

$$y = x^3 - 12x - 12,$$

and sketch the graph.

Find the set of values of k for which the equation

$$x^3 - 12x - 12 = k$$

has more than one real solution.

Taking a suitable first approximation, use the Newton-Raphson method to obtain a sequence of approximations for the positive root of the equation.

$$x^3 - 12x - 12 = 0$$

Your sequence should terminate when you consider that you have found the root correct to three decimal places.

Cambridge June 1987

Suggested solution

$$y = x^3 - 12x - 12$$

To find maxima or minima differentiate y with respect to x and equate to zero.

$$\frac{dy}{dx} = 3x^2 - 12 = 0$$

ie $\quad 3(x^2 - 4) = 0$

either $3 = 0!$ $\quad$ or $\quad x^2 - 4 = 0$

$$x = \pm 2$$

second differential with respect to x

$$\frac{d^2y}{dx^2} = 6x$$

when $x = 2$ ..... $\dfrac{d^2y}{dx^2} = +12$ ..... minimum

when $x = -2$ ..... $\dfrac{d^2y}{dx^2} = -12$ ..... maximum

values for y

when $x = +2$ ..... $y = (2)^3 - 12(2) - 12 \quad = -28$

when $x = -2$ ..... $y = (-2)^3 - 12(-2) - 12 \quad = +4$

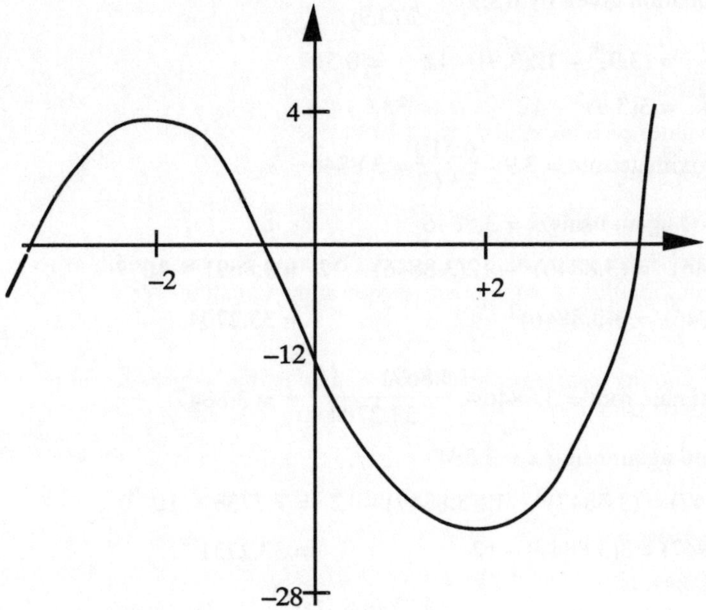

$$x^3 - 12x - 12 = k$$

ie $$x^3 - 12x - (12 + k) = 0$$

As this is a cubic equation, the possible roots are

i) 1 real and 2 imaginary

ii) all roots are real and 2 roots are equal

iii) all roots are real and unequal

The latter two are only possible if the curve lies between the maximum and minimum values +4 and –28 ie

$$-28 < k < +4$$

First approximation for positive root:

try $x = 3$ ..... $y = (3)^3 - 12(3) - 12 \Rightarrow -21$

try $x = 4$ ..... $y = (4)^3 - 12(4) - 12 \Rightarrow +4$

The positive root lies between $x = 3$ and $x = 4$ and closer to 4, as a first approximation choose $x = 3.9$.

Using Newton-Raphson method

second approximation given by $f(3.9) - \dfrac{f(3.9)}{f'(3.9)}$

where $f(3.9) = (3.9)^3 - 12(3.9) - 12 = 0.519$

and $f'(3.9) = 3(3.9)^2 - 12 = 33.63$

ie second approximate root $= 3.9 - \dfrac{0.519}{33.63} = 3.8846$

apply the method again using $x = 3.8846$

ie $f(3.8846) = (3.8846)^3 - 12(3.8846) - 12 = 3.8691 \times 10^{-3}$

$f'(3.8846) = 3(3.8846)^2 - 12 = 33.2704$

ie third approximate root $= 3.8846 - \dfrac{3.8691 \times 10^{-3}}{33.2704} = 3.8847$

apply the method again using $x = 3.8847$

ie $f(3.8847) = (3.8847)^3 - 12(3.8847) - 12 = 7.7738 \times 10^{-3}$

$f'(3.8847) = 3(3.8847) - 12 = 33.2731$

ie fourth approximate root $= 3.8847 - \dfrac{7.7738 \times 10^{-3}}{33.2731} = 3.8849$

These approximate roots are consistent to three decimal places

ie approximate root to 3 decimal places = 3.885

Question 2

Show by induction, or otherwise, that

$$1 \times 3 + 3 \times 5 + 5 \times 7 + \ldots (2n-1) \times (2n+1) = \frac{1}{6}(2n-1)(2n+1)(2n+3) + \frac{1}{2}$$

Oxford June 1983

Suggested solution

Suppose the summation is true for a value $n = r$

The next term on the left hand side is given by $n = r + 1$

ie $\quad [2(r+1) - 1][2(r+1) + 1] = (2r+1)(2r+3)$

Add this new term to both sides of the equation and right hand side becomes

$$\text{RHS} = \frac{1}{6}(2r-1)(2r+1)(2r+3) + \frac{1}{2} + (2r+1)(2r+3)$$

$$= \frac{1}{6}(2r+1)(2r+3)\,[(2r-1) + 6] + \frac{1}{2}$$

$$= \frac{1}{6}(2r+1)(2r+3)(2r+5) + \frac{1}{2}$$

The right hand side now has the same format as that of the original expression except that r has now become r + 1

ie $\quad$ if the original expression was true for $n = r$ it is also true for $n = r + 1$.

To test whether the original expression is true for any value of n, let n have the value 1.

Left hand side $= 1 \times 3 = 3$

$$\text{Right hand side} = \frac{1}{6}[2(1) - 1][2(1) + 1][2(1) + 3] + \frac{1}{2}$$

$$= \frac{5}{2} + \frac{1}{2} = 3$$

ie $\quad$ the original expression in true for $n = 1$ and is therefore true for any value of n.

Question 3

Obtain the expansion in ascending powers of x of the function

$$f(x) = \ln\left[\frac{(1-3x)^2}{(1+2x)}\right]$$

Give the values of the coefficients of the powers of x up to and including x^3, and give an expression for the coefficient of x^n. For what range of values of x is this expansion valid?

Oxford June 1983

Suggested solution

Standard expansion:

$$\ln(1+z) = z - \frac{z^2}{2} + \frac{z^3}{3} - \frac{z^4}{4} + \ldots\ldots \pm \frac{z^n}{n} \qquad \text{if } -1 \le z \le +1$$

$$\ln\left[\frac{(1-3x)^2}{(1+2x)}\right] = 2.\ln(1-3x) - \ln(1+2x)$$

expand $\ln(1-3x)$ by putting $z = -3x$

ie $\quad \ln(1-3x) = (-3x) - \frac{(-3x)^2}{2} + \frac{(-3x)^3}{3} \ldots\ldots \pm \frac{(-3x)^n}{n}$

$$= -3x - \frac{9x^2}{2} - 9x^3 \ldots\ldots \pm \frac{(-3x)^n}{n}$$

$\therefore \quad 2.\ln(1-3x) \quad = -6x - 9x^2 - 18x^3 \ldots\ldots \pm \frac{2(-3x)^n}{n}$

expand $\ln(1+2x)$ by putting $z = 2x$

ie $\quad \ln(1+2x) = 2x - \frac{(2x)^2}{2} + \frac{(2x)^3}{3} \ldots\ldots \pm \frac{(2x)^n}{n}$

$$= 2x - 2x^2 + \frac{8x^3}{3} \ldots\ldots \pm \frac{(2x)^n}{n}$$

hence $\quad 2.\ln(1-3x) - \ln(1+2x)$

$$= \left[-6x - 9x^2 - 18x^3 \ldots\ldots \pm \frac{2(-3x)^n}{n}\right] - \left[2x - 2x^2 + \frac{8x^3}{3} \ldots\ldots \pm \frac{(2x)^n}{n}\right]$$

$$= -8x - 7x^2 - \frac{62}{3}x^3 \ldots\ldots \pm \left[\frac{2(-3x)^n}{n} - \frac{(2x)^n}{n}\right]$$

thus, coefficients are: $x \Rightarrow -8$; $x^2 \Rightarrow -7$; $x^3 \Rightarrow \frac{-62}{3}$

General coefficient of $x^n \Rightarrow \pm \frac{2}{n}[(-3)^n - (2)^{n-1}]$

choosing + sign when n is an odd number of terms

– sign when n is an even number of terms

This can be checked:

if n = 1 then $\Rightarrow + \frac{2}{1}[(-3)^1 - 2^0] \Rightarrow -8$

if n = 2 then $\Rightarrow - \frac{2}{2}[(-3)^2 - 2^1] \Rightarrow -7$

if n = 3 then $\Rightarrow + \frac{2}{3}[(-3)^3 - 2^2] \Rightarrow - \frac{62}{3}$

Range of values of x for which the expansion is valid

expansion of ln(1 + z) is valid if $-1 < z \le +1$

hence for $\ln(1 - 3x)$ ..... $-1 < -3x \le +1$

$\Rightarrow +1 > 3x \ge -1$

$\Rightarrow +\frac{1}{3} > x \ge -\frac{1}{3}$

and for $\ln(1 + 2x)$ ..... $-1 < 2x \le +1$

$\Rightarrow -\frac{1}{2} < x \le +\frac{1}{2}$

ie

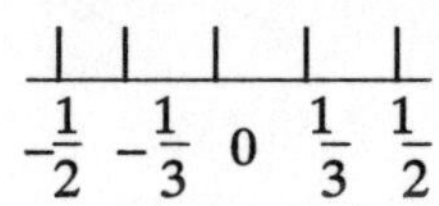

Range of x to satisfy both expansions is $-\frac{1}{2} < x \le +\frac{1}{2}$

8 CURVES, POLAR COORDINATES AND PARAMETERS

8.1 Key formulae and ideas

1. Curve sketching ..... Cartesian coordinates ..... values to be determined

a) points of intersection on axes

..... put $x = 0$, calculate y

..... put $y = 0$, calculate x

b) turning points of curve ..... put $\frac{dy}{dx} = 0$ to find x value

c) nature of turning points ..... find $\frac{d^2y}{dx^2}$ ..... insert x value

if $\frac{d^2y}{dx^2}$ ..... is negative ..... maximum

is positive ..... minimum

if $\frac{d^2y}{dx^2} = 0$ ..... point of inflection $\frac{d^2y}{dx^2}$ changes

sign as curve passes through point

but $\frac{dy}{dx}$ does not change sign

d) symmetry of curve

e) equations of asymptotes ..... let x and y in turn $\Rightarrow \pm\infty$ ..... curve approximates to a straight line

f) regions where the curve cannot exist

2 **Polar coordinates ..... position of a point is determined by distance (r) from origin and angle (θ) subtended to the *x*-axis.**

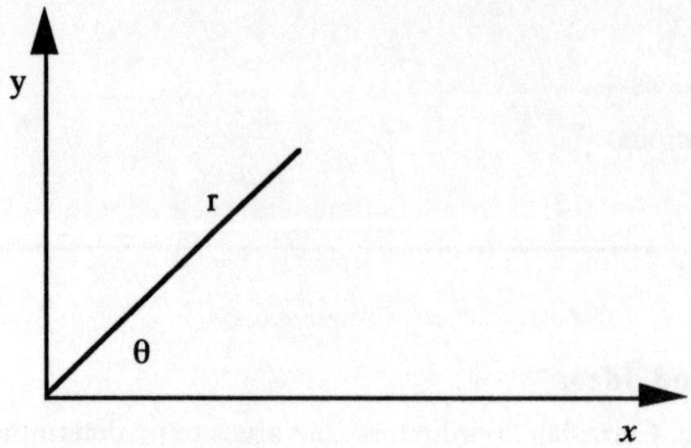

relation between Cartesian and polar coordinates

$y = r\sin\theta$ $\quad x = r.\cos\theta$ $\qquad \tan\theta = \frac{y}{x}$ $\qquad r = \sqrt{(x^2 + y^2)}$

area under curve $= \frac{1}{2}\int_{\alpha}^{\beta} r^2\theta \, d\theta$

8.1 Questions and solutions

Question 1

For the curve $y = (3 - 2x)(x^2 - 36)$, prove that $\frac{dy}{dx} > 0$

if, and only if, $-3 < x < 4$.

Sketch the curve $y = (3 - 2x)(x^2 - 36)$, clearly labelling with their coordinates

a) the turning points

b) the points where the curve intersects the coordinate axes.

Hence show, in a separate sketch, the general shape of the curve

$y^2 = (3 - 2x)(x^2 - 36)$.

AEB June 1984

Suggested solution

Intercepts on axes

put $x = 0$ $\quad y = (3 - 0)(0 - 36) = -108$

put $y = 0$ $\quad 0 = (3 - 2x)(x^2 - 36)$

either $3 - 2x = 0$ $\quad$ or $x^2 - 36 = 0$

$x = \frac{+3}{2}$ $\quad$ $x = \pm 6$

Turning points ...... treat as differentiation of a product

$$\frac{dy}{dx} = (3 - 2x)(2x) + (x^2 - 36)(-2)$$

$$= -6x^2 + 6x + 72 = -6(x^2 - x + 12)$$

hence $\frac{dy}{dx} = 0$ if $\quad x^2 - x - 12 = 0$

$$(x - 4)(x + 3) = 0$$

if $\frac{dy}{dx} > 0$ then both brackets are positive or both brackets are negative

ie $\quad -3 > x > +4$

The turning points are given by $x = -3$ or $x = +4$

if $x = -3$ ..... $y = [3 - 2(-3)][(-3)^2 - 36] \quad = - 243$

if $x = +4$ ..... $y = [3 - 2(4)][4^2 - 36] \quad = + 100$

nature of turning points

$$\frac{d^2y}{dx^2} = - 12x + 6$$

if $x = -3$ $\quad \dfrac{d^2y}{dx^2} = - 12(-3) + 6 = + 42$ ..... minimum

if $x = +4$ $\quad \dfrac{d^2y}{dx^2} = - 12(4) + 6 = - 18$ ..... maximum

$y = (3 - 2x)(x^2 - 36)$

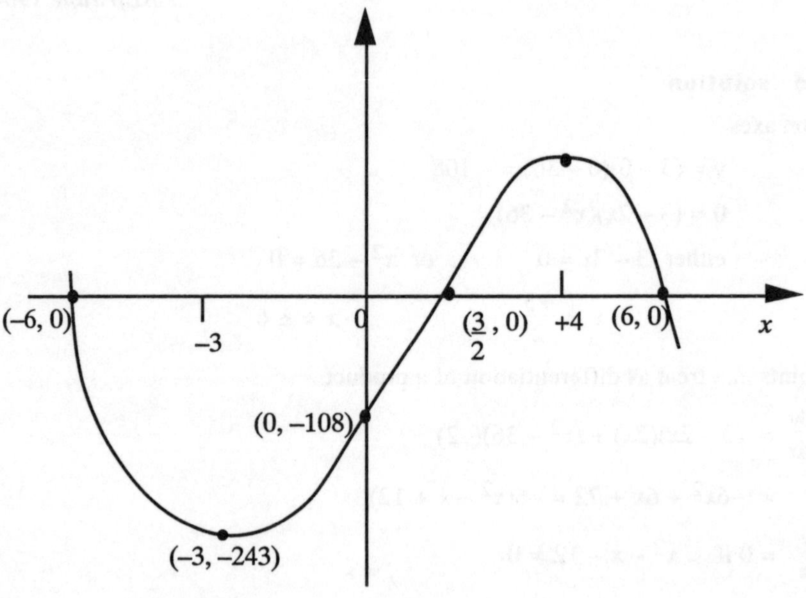

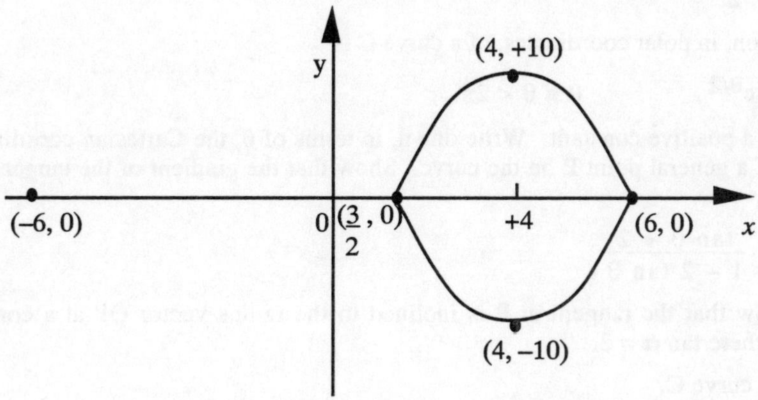

If $x = 0$, $y^2 = -108$ ..... $y = \sqrt{-108}$ is an imaginary number

$y = 0$, $y^2 = 0$ $x = \pm 6$ or $x = (3/2)$

$x = 4$, $y^2 = 100$ $y = \pm 10$

$x = -3$, $y^2 = -243$ ..... $y = \sqrt{-243}$ is an imaginary number

Question 2

The equation, in polar coordinates, of a curve C is

$$r = a\,e^{\theta/2}, \qquad 0 \le \theta < 2\pi$$

where a is a positive constant. Write down, in terms of θ, the Cartesian coordinates, x and y, of a general point P on the curve. Show that the gradient of the tangent at P is given by

$$\frac{dy}{dx} = \frac{\tan\theta + 2}{1 - 2\tan\theta}$$

Hence show that the tangent at P is inclined to the radius vector OP at a constant angle α, where tan α = 2.

Sketch the curve C.

[An accurate graph is not required but you should show clearly the range of values of r.]

Find the area of the region traced out by the radius vector OP as P moves on C from the point where θ = 0 to the point where θ = π.

Cambridge June 1989

Suggested solution

In polar coordinates

$$r = ae^{\frac{1}{2}\theta}$$

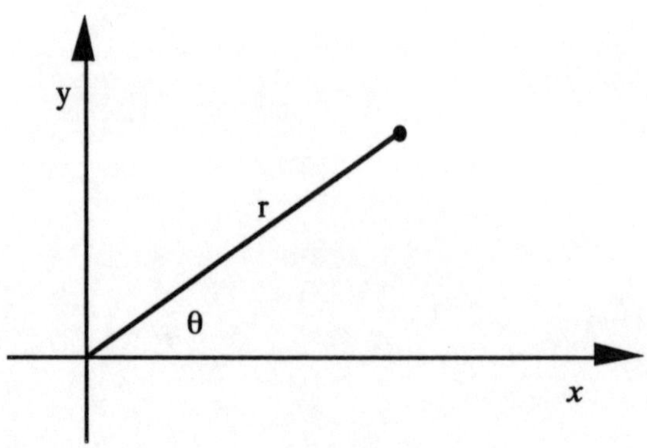

and $x = r\cos\theta$

$y = r\sin\theta$

$$\frac{dy}{dx} = \frac{dy}{d\theta} \bullet \frac{d\theta}{dx}$$

To find $\frac{dy}{d\theta}$ $\quad y = ae^{\frac{1}{2}\theta} \sin\theta$ ..... a = constant

$$\frac{dy}{d\theta} = a\left[e^{\frac{1}{2}\theta} \cos\theta + \sin\theta \,.\, \left(\frac{1}{2}\right) . \, e^{\frac{1}{2}\theta}\right]$$

$$= \frac{a.e^{\frac{1}{2}\theta}}{2} [2\cos\theta + \sin\theta]$$

To find $\frac{dx}{d\theta}$ $\quad x = ae^{\frac{1}{2}\theta} \cos\theta$

$$\frac{dx}{d\theta} = a\left[e^{\frac{1}{2}\theta} (-\sin\theta) + \cos\theta \,.\, \left(\frac{1}{2}\right).e^{\frac{1}{2}\theta}\right]$$

$$= \frac{a.e^{\frac{1}{2}\theta}}{2} [\cos\theta - 2\sin\theta]$$

$$\frac{dy}{dx} = \frac{ae^{\frac{1}{2}\theta}}{2} [2\cos\theta + \sin\theta] . \frac{2}{ae^{\frac{1}{2}\theta} [\cos\theta - 2\sin\theta]}$$

$$= \frac{2\cos\theta + \sin\theta}{\cos\theta - 2\sin\theta}$$ divide numerator and denominator by $\cos\theta$

$$= \frac{2 + \tan\theta}{1 - 2\tan\theta}$$

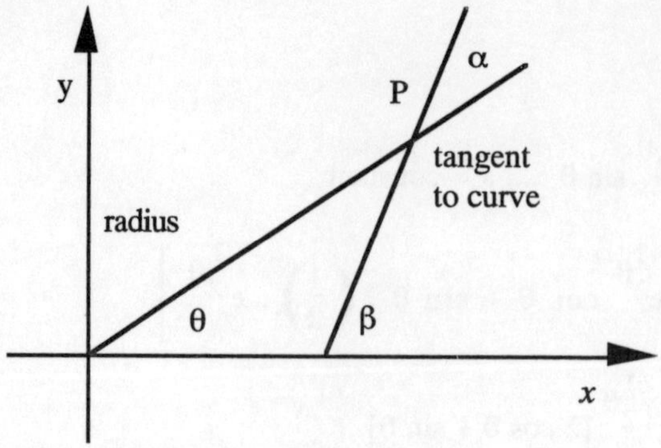

Let tangent to the curve at P be inclined to the x–axis at angle β.

Angle between tangent and radius = α = (β – θ)

$$\tan\alpha = \tan(\beta - \theta) = \frac{\tan\beta - \tan\theta}{1 + \tan\theta.\tan\beta}$$

but $\tan\beta = \dfrac{dy}{dx}$

$$\tan\alpha = \frac{\left[\dfrac{2 + \tan\theta}{1 - 2\tan\theta}\right] - \tan\theta}{1 + \left[\dfrac{2 + \tan\theta}{1 - 2\tan\theta}\right]\tan\theta}$$

$$= \frac{[2 + \tan\theta - \tan\theta\,(1 - 2\tan\theta)][1 - 2\tan\theta]}{[1 - 2\tan\theta][1 - 2\tan\theta + 2\tan\theta + \tan^2\theta]}$$

$$= \frac{2 + 2\tan^2\theta}{1 + \tan^2\theta} = 2$$

To sketch the curve ..... this should be done in terms of r and θ ..... values need to be calculated.

		$r = ae^{\frac{1}{2}\theta}$	θ in radians
when	$\theta = 0$	$r = ae^0$	$= a$
	$\theta = \pi/4$	$r = ae^{\pi/8}$	$= 1.481\ a$
	$\theta = \pi/2$	$r = ae^{\pi/4}$	$= 2.193\ a$
	$\theta = 3\pi/4$	$r = ae^{3\pi/8}$	$= 3.248\ a$

$\theta = \pi$	$r = ae^{\pi/2}$	$= 4.810\ a$
$\theta = 5\pi/4$	$r = ae^{5\pi/8}$	$= 7.214\ a$
$\theta = 3\pi/2$	$r = ae^{3\pi/4}$	$= 10.55\ a$
$\theta = 7\pi/4$	$r = ae^{7\pi/8}$	$= 15.625\ a$
$\theta = 2\pi$	$r = ae^{\pi}$	$= 23.141\ a$

The sketch diagram, although not required to be drawn to scale, should look realistic and it is better to plot it with a reasonable degree of accuracy.

Since 'a' is a constant it is sensible to give it a nominal value eg let a = 1.

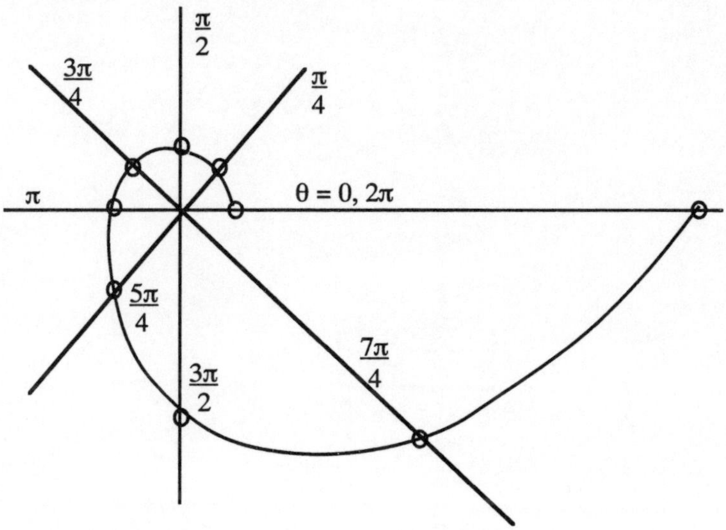

Area under curve as θ varies from 0 to π.

In polar coordinates

$$\text{area under curve} = \frac{1}{2}\int_{\alpha}^{\beta} r^2\, d\theta$$

ie $$\text{area} = \frac{1}{2}\int_{0}^{\pi} (ae^{\theta/2})^2\, d\theta$$

$$= \frac{a^2}{2} \int_0^{\pi} e^{\theta}\, d\theta = \frac{a^2}{2} [e^{\theta}]_0^{\pi}$$

$$= \frac{a^2}{2} [e^{\pi} - e^0] = \frac{a^2}{2} [23.141 - 1]$$

$$= 11.07\, a^2$$

Question 3

Referred to an origin 0 and coordinate axes 0x and 0y, a curve is given by

$$x = \sec t + \tan t, \qquad y = \operatorname{cosec} t + \cot t, \qquad 0 < t < \pi/2$$

where t is a parameter.

Prove that $\dfrac{dy}{dx} = \dfrac{-(1 - \sin t)}{(1 - \cos t)}$

Show that the normal to the curve at the point S, where $t = \tan^{-1}(3/4)$, has equation $x - 2y + 4 = 0$. Find an equation of the normal to the curve at the point T, where $t = \tan^{-1}(4/3)$. These normals meet at the point N. Find the coordinates of the point N.

Hence calculate

a) the area of the triangle SNT

b) the tangent of the angle SNT.

AEB June 1986

Suggested solution

$$\frac{dy}{dx} = \frac{dy}{dt} \cdot \frac{dt}{dx}$$

To find $\dfrac{dy}{dt}$

$$y = \operatorname{cosec} t + \cot t$$

$$\frac{dy}{dt} = -\operatorname{cosec}(t).\cot(t) - \operatorname{cosec}^2(t)$$

$$= -\frac{\cos(t)}{\sin^2(t)} - \frac{1}{\sin^2(t)}$$

$$= -\frac{[\cos(t) + 1]}{[1-\cos(t)][1+\cos(t)]}$$

$$= -\frac{1}{[1 - \cos(t)]}$$

To find $\dfrac{dx}{dt}$

$$x = \sec(t) + \tan(t)$$

$$\frac{dx}{dt} = \sec(t).\tan(t) + \sec^2(t)$$

$$= \frac{\sin(t)}{\cos^2(t)} + \frac{1}{\cos^2(t)}$$

$$= \frac{[\sin(t) + 1]}{[1 - \sin(t)][1 + \sin(t)]}$$

$$= \frac{1}{[1 - \sin(t)]}$$

hence $\frac{dy}{dx} = \frac{dy}{dt} \bullet \frac{dt}{dx} \quad = -\frac{1}{[1 - \cos(t)]} \bullet \frac{[1 - \sin(t)]}{1}$

$$= -\frac{[1 - \sin(t)]}{[1 - \cos(t)]}$$

Normal to curve at point S

gradient of normal × gradient of tangent = –1

gradient of tangent $= \frac{dy}{dx} = -\frac{[1 - \sin(t)]}{[1 - \cos(t)]}$

$t = \tan^{-1}(3/4)$

ie $\quad \sin(t) = (3/5)$

$\cos(t) = (4/5)$

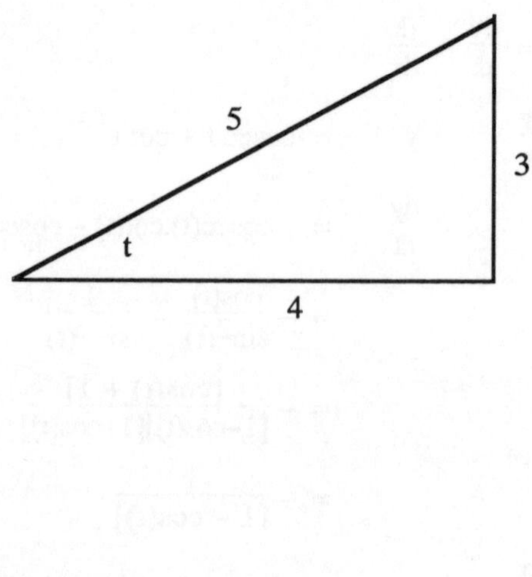

$$\frac{dy}{dx} = -\frac{\left[1 - \frac{3}{5}\right]}{\left[1 - \frac{4}{5}\right]} = -2$$

gradient of normal $= +\frac{1}{2}$

At point S $\quad y = \text{cosec}(t) + \cot(t) \quad = \frac{5}{3} + \frac{4}{3} = 3$

$x = \sec(t) + \tan(t) \quad = \frac{5}{4} + \frac{3}{4} = 2$

Equation of normal is given by $y = mx + c$

ie $\quad 3 = \frac{1}{2}(2) + c \quad$ thus $c = 2$

equation of normal at point S $\quad y = \frac{1}{2}x + 2$

$$x - 2y + 4 = 0$$

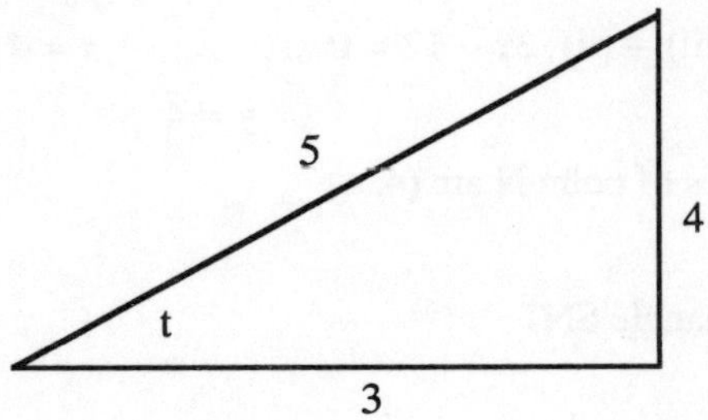

Normal at point T

$t = \tan^{-1}(4/3)$

$\sin(t) = (4/5)$

$\cos(t) = (3/5)$

$$\text{gradient of tangent at T} = -\frac{\left[1 - \frac{4}{5}\right]}{\left[1 - \frac{3}{5}\right]} = -\frac{1}{2}$$

$$\text{gradient of normal at T} = -\frac{1}{-\frac{1}{2}} = +2$$

coordinates of T $\quad x = \sec(t) + \tan(t) \quad = \frac{5}{3} + \frac{4}{3} = 3$

$\quad y = \operatorname{cosec}(t) + \cot(t) \quad = \frac{5}{4} + \frac{3}{4} = 2$

equation of normal at T $\quad y = mx + k$

$\quad 2 = 2(3) + k \;.....\; k = -4$

hence $\quad y = 2x - 4$

ie $\quad 2x - y - 4 = 0$

Coordinates of point N are given by the intersection of the line SN

$x - 2y + 4 = 0$ ..... (i)

and line TN

$2x - y - 4 = 0$ ..... (ii)

multiply equation (ii) by 2

$4x - 2y - 8 = 0$ ..... (iii)

subtract (iii) – (ii) $3x - 12 = 0$ ..... $x = 4$

hence $y = 4$

coordinates of point N are (4, 4)

Area of triangle SNT

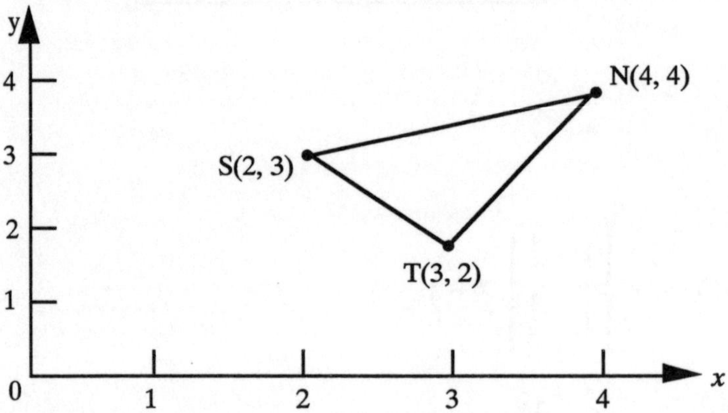

in general the area of a triangle is given by

$$\pm \frac{1}{2} \begin{vmatrix} x_1 & y_1 & 1 \\ x_2 & y_2 & 1 \\ x_3 & y_3 & 1 \end{vmatrix}$$

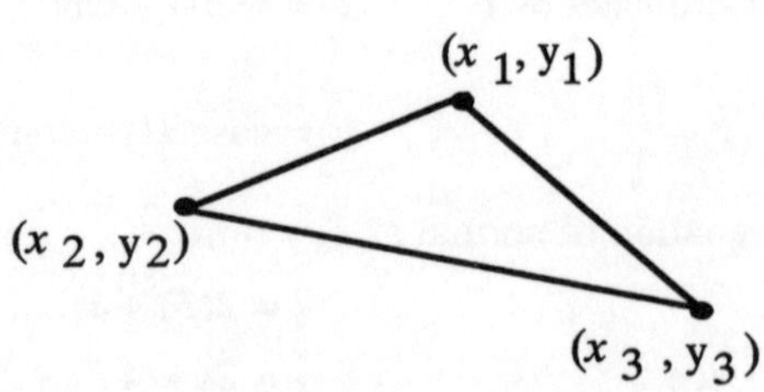

$= \pm \frac{1}{2}[x_1y_2 + x_2y_3 + x_3y_1 - x_3y_2 - x_2y_1 - x_1y_3]$

$= \pm \frac{1}{2}[(4 \times 3) + (2 \times 2) + (3 \times 4) - (3 \times 3) - (2 \times 4) - (4 \times 2)]$

$= \pm \frac{3}{2}$

area of triangle SNT $= \frac{3}{2}$ ..... choosing positive value

Angle $S\hat{N}T$ $\quad \tan S\hat{N}T = \frac{m_1 - m_2}{1 + m_1.m_2} = \frac{2 - \frac{1}{2}}{1 + 2\left(\frac{1}{2}\right)} = \frac{3}{4}$

ie $\quad \tan S\hat{N}T = \frac{3}{4}$

9 COMPLEX NUMBERS

9.1 Key formulae and ideas

1. Rational numbers

..... all simple numbers dealt with in elementary arithmetic ..... positive and negative integers, fractions, which are ratios of integers, and zero which is the number separating positive and negative numbers.

Irrational numbers ..... such as $\sqrt{2}$

2. Real numbers

..... the limit of a sequence of rational numbers a_n as the integer n increases indefinitely ..... a completely ordered set.

3. Complex number (z) = $(x, y) = (x + iy)$

where i is such that $i^2 = -1$

Argand diagram ..... a means of representing complex numbers graphically

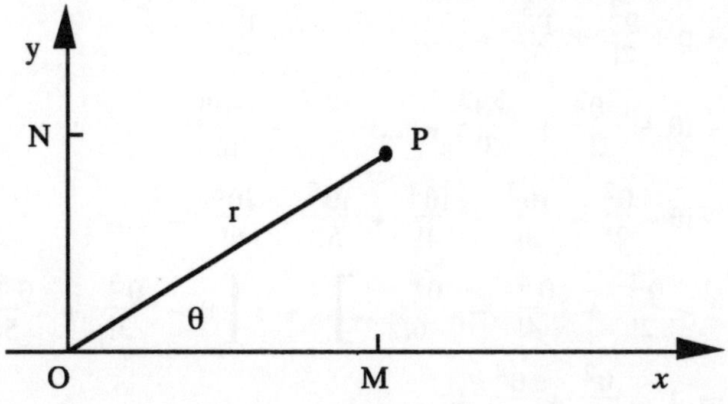

OM = $x = r\cos\theta$; ON = $y = r\sin\theta$

Absolute value or MODULUS $|z|$ = length of vector OP (r)

$$= \sqrt{x^2 + y^2}$$

ARGUMENT or Amplitude (θ) = angle subtended by vector OP to x direction

such that $y = r \sin\theta$

] ..... $\tan\theta = \frac{y}{x}$

and $x = r \cos\theta$

Two complex numbers $z_1 = x_1 + iy_1$ and $z_2 = x_2 + iy_2$ are equal only if

$x_1 = x_2$ AND $y_1 = y_2$.

A complex number = 0 only if $x = 0$ AND $y = 0$

4. Complex numbers obey rules of addition, subtraction and multiplication but cannot be divided without first rationalising the denominator ie by multiplying the denominator by its conjugate.

$$\text{eg } \frac{x + iy}{a + ib} \Rightarrow \frac{(x + iy)(a - ib)}{(a + ib)(a - ib)} \Rightarrow \frac{(x + iy)(a - ib)}{a^2 + b^2} \qquad \text{as } i^2 = -1$$

5. Polar coordinates

$$z = x + iy = r \cos\theta + i.r \cos\theta = r(\cos\theta + i \sin\theta)$$

6. De Moivre's theorem

$$z^n = (x + iy)^n = r^n(\cos\theta + i \sin\theta)^n = r^n(\cos n\theta + i \sin n\theta)$$

7. Exponential form of a complex number

$$e^p = 1 + p + \frac{p^2}{2!} + \frac{p^3}{3!} + \ldots\ldots + \frac{p^n}{n!}$$

$$e^{i\theta} = 1 + i\theta + \frac{i^2\theta^2}{2!} + \frac{i^3\theta^3}{3!} + \ldots\ldots + \frac{i^n\theta^n}{n!}$$

$$= 1 + i\theta - \frac{\theta^2}{2!} - \frac{i\theta^3}{3!} + \frac{i\theta^4}{4!} + \frac{i\theta^5}{5!} - \frac{i\theta^6}{6!} - \ldots\ldots$$

$$= \left[1 - \frac{\theta^2}{2!} + \frac{\theta^4}{4!} - \frac{\theta^6}{6!}\ldots\right] + i\left[\theta - \frac{\theta^3}{3!} + \frac{\theta^5}{5!} - \ldots\right]$$

$$\text{but } \cos\theta = 1 - \frac{\theta^2}{2!} + \frac{\theta^4}{4!} - \ldots\ldots$$

$$\text{and } \sin\theta = \theta - \frac{\theta^3}{3!} + \frac{\theta^5}{5!} - \ldots\ldots$$

hence $e^{i\theta} = (\cos\theta + i\sin\theta)$

ie $z = (x + iy) = r(\cos\theta + i\sin\theta) = r\,e^{i\theta}$

9.2 Questions and solutions

Question 1

a) Given that p and q are real, find the possible values of p and q such that

$$(p + iq)^2 = -15 - 8i.$$

Hence solve the quadratic equation

$$z^2 - (3 - 2i)z + (5 - i) = 0$$

Why are the two roots not complex conjugates?

b) Given that $1 - \sqrt{3}i$ is a root of the equation

$$z^4 - 4z^3 - 2z^2 + 12z + k = 0,$$

where k is a real constant, write down another root of the equation and hence by forming a quadratic factor, or otherwise, determine the value of k and the other roots.

Oxford June 1988

Suggested solution

a)
$$(p + iq)^2 = -15 - 8i$$
$$p^2 + 2pqi - q^2 = -15 - 8i$$

equate real and imaginary parts

ie $p^2 - q^2 = -15$ and $2pq = -8$

$$p = -\frac{4}{q}$$

substitute for p $\left[-\frac{4}{q}\right]^2 - q^2 = -15$

$$q^4 - 15q^2 - 16 = 0$$
$$(q^2 - 16)(q^2 + 1) = 0$$

either $q^2 - 16 = 0$ ie $q^2 = 16$ ie $q = \pm 4$

or $q^2 + 1 = 0$ ie $q^2 = -1$ ie $q = i$

Hence if $q = +4$ $p = -1$

$q = -4$ $p = +1$

Also if $q = i$ then $p^2 - q^2 = p^2 + 1 = -15$

ie $p^2 = -16$

and $p = 4i$ ..... not a permitted solution as both p and q are real.

$$z^2 - (3 - 2i)z + (5 - i) = 0$$

ie $$z = \frac{+(3 - 2i) \pm \sqrt{(3 - 2i)^2 - 4(1)(5 - i)}}{2(1)}$$

$$z = \frac{+(3 - 2i) \pm \sqrt{(-15 - 8i)}}{2}$$

if $z =$ complex number $= (p + iq)$

then $$p + iq = \frac{(3 - 2i) \pm (p + iq)}{2}$$

either $3(p + iq) = 3 - 2i$ or $p + iq = 3 - 2i$

ie $3p = 3$ and $3q = -2$ $p = 3$ and $q = -2$

$p = 1$ and $q = \dfrac{-2}{3}$

solutions are: $z_1 = 1 - \dfrac{2}{3}i$ and $z_2 = 3 - 2i$

Two complex numbers are said to be conjugate when they differ only in the sign of the imaginary part.

b) $$z^4 - 4z^3 - 2z^2 + 12z + k = 0$$

given that one root is $z = 1 - \sqrt{3}$

ie one factor is $[z - (1 - \sqrt{3})]$

a second root is the conjugate $z = 1 + \sqrt{3}$

ie a second factor is $[z - (1 + \sqrt{3})]$

multiply these two factors to form a quadratic factor

ie $[z - (1 - \sqrt{3})][z - (1 + \sqrt{3})] = z^2 - 2z + 4$

divide the original expression by this quadratic factor to obtain a second quadratic factor which will lead to k.

$$\begin{array}{r|l} & z^2 - 2z - 10 \\ \hline z^2 - 2z + 4 & z^4 - 4z^3 - 2z^2 + 12z + k \\ & z^4 - 2z^3 + 4z^2 \\ \hline & \quad - 2z^3 - 6z^2 + 12z \\ & \quad - 2z^3 + 4z^2 - 8z \\ \hline & \qquad - 10z^2 + 20z + k \\ & \qquad - 10z^2 + 20z - 40 \\ \hline & \qquad\qquad k - 40 \end{array}$$

but for factors the remainder $k - 40 = 0$

ie $k = 40$

also the quadratic factor $z^2 - 2z - 10$ will yield two further roots of z

ie $z^2 - 2z - 10 = 0$

$$z = \frac{-(-2) \pm \sqrt{(-2)^2 - 4(1)(-10)}}{2(1)}$$

$$= \frac{2 \pm \sqrt{44}}{2} = 1 \pm \sqrt{11}$$

The remaining roots are either $z = 1 + \sqrt{11}$

or $z = 1 - \sqrt{11}$

Question 2

In an Argand diagram the points A, B and C represent the complex numbers a, 6 + 8i and c respectively, and 0 is the origin. OABC is a square described in an anticlockwise sense. Give a reason why c = ia and a reason why a + c = 6 + 8i.

Find, by calculation, a and c.

On a single, clearly labelled sketch, show the points O, A, B, C and the loci

i) $|z - 3 - 4i| = 5$,

ii) $|z| = |z - 6 - 8i|$,

iii) $\arg(z - a) - \arg z = \frac{1}{2}\pi$

Cambridge June 1989

Suggested solution

c = ia ..... complex number a represents side OA of a square ..... complex number c represents the adjacent side OC of the square where OA and OC are perpendicular.

From diagram

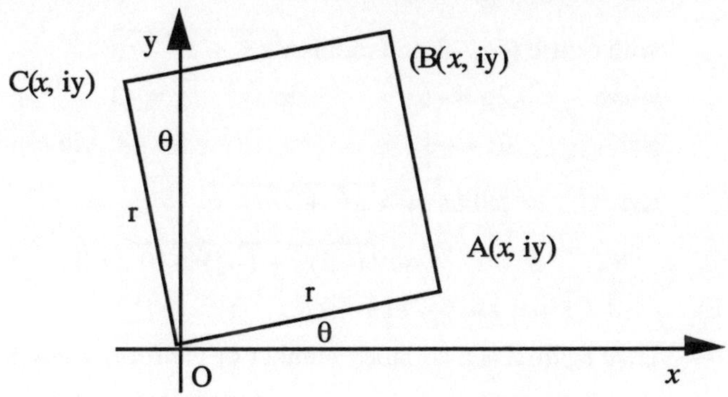

$$a = x + iy$$
$$= r(\cos\theta + i\sin\theta)$$
$$c = -x + iy$$
$$= -r(\sin\theta - i\cos\theta)$$
$$= ir(\cos\theta + i\sin\theta)$$
$$= ia$$

From diagram a + c

$\Rightarrow \underline{OA} + \underline{AB}$

$\Rightarrow \underline{OB} = 6 + 8i$

to calculate a and c

a + c = 6 + 8i and c = ia

substitute for c

a + ia = 6 + 8i and if a = x + iy

$$x + iy + i(x + iy) = 6 + 8i$$
$$(x - y) + i(x + y) = 6 + 8i$$

equate coefficients $x - y = 6$

$x + y = 8$

$$2x = 14 \; ..... \; x = 7, \;\; y = 1$$

Hence $a = x + iy = 7 + i(1) = 7 + i$

and $c = i(x + iy) = i(7 + i) = -1 + 7i$

i) $|z - 3 - 4i| = 5$

z is a complex number of form $z = x + iy$

hence $|z - 3 - 4i| = |x + iy - 3 - 4i|$

$= |(x - 3) + i(y - 4)|$

$= \sqrt{(x - 3)^2 + (y - 4)^2} = 5$

$(x - 3)^2 + (y - 4)^2 = 5^2$

$x^2 - 6x + 9 + y^2 - 8y + 16 = 25$

$x^2 + y^2 - 6x - 8y = 0$

This equation has the standard form of a circle

$x^2 + y^2 + 2gx + 2fy + c = 0$

with centre $(-g, -f)$ and radius $\sqrt{g^2 + f^2 - c}$

where $2g = -6$ ie $g = -3$

and $2f = -8$ $f = -4$ ie centre $(3, 4)$

and radius $= \sqrt{g^2 + f^2 - c}$

$= \sqrt{(-3)^2 + (-4)^2 - 0} = 5$

ii) $|z| = |z - 6 - 8i|$

Here again z is a complex number of the form $x + iy$ in which case

$|z| = \sqrt{x^2 + y^2}$

and $|z - 6 - 8i| = |x + iy - 6 - 8i|$

$= |(x - 6) + i(y - 8)|$

$= \sqrt{(x - 6)^2 + (y - 8)^2}$

hence $\sqrt{x^2 + y^2} = \sqrt{(x - 6)^2 + (y - 8)^2}$

square $x^2 + y^2 = x^2 - 12x + 36 + y^2 - 16y + 64$

ie $12x + 16y = 100$

$3x + 4y = 25$

$$y = -\frac{3}{4}x + \frac{25}{4}$$

Compare this with the standard equation of a straight line $y = mx + c$

intercept on y-axis $= \frac{25}{4}$

intercept on x-axis $= \frac{25}{3}$

gradient = $\tan\theta = -\frac{3}{4}$

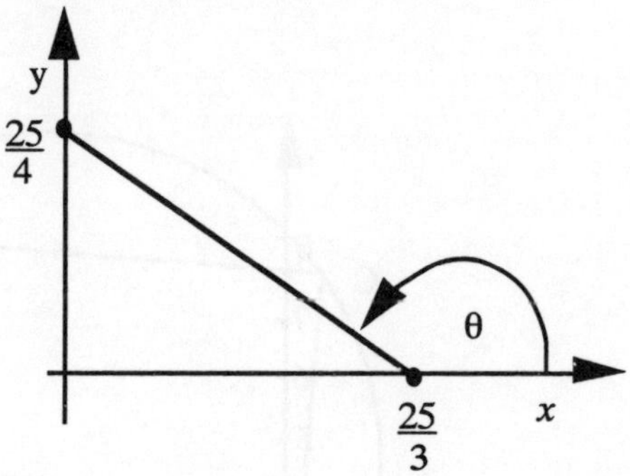

iii) $\arg(z - a) - \arg(z) = \frac{\pi}{2}$

again let $z = x + iy$ and $a = 7 + i$

hence $\arg[(x + iy) - (7 + i)] - \arg[x + iy] = \frac{\pi}{2}$

$$\arg[(x - 7) + i(y - 1)] - \arg[x + iy] = \frac{\pi}{2}$$

ie $\tan^{-1}\left[\frac{y - 1}{x - 7}\right] - \tan^{-1}\left[\frac{y}{x}\right] = \frac{\pi}{2}$

let $\alpha = \tan^{-1}\left[\frac{y - 1}{x - 7}\right]$ and $\beta = \left[\frac{y}{x}\right]$

ie $\alpha - \beta = \frac{1}{2}\pi$ ..... $\alpha = \frac{1}{2}\pi + \beta$

$$\tan\alpha = -\frac{1}{\tan\beta}$$

hence $\frac{y - 1}{x - 7} = -\frac{x}{y}$

$$y^2 - y = -x^2 + 7x$$

ie $x^2 + y^2 - 7x - y = 0$

Again this is the equation of a circle where

$2g = -7$ ie $g = -\frac{7}{2}$ and $2f = -1$ ie $f = -\frac{1}{2}$

Centre of circle $= (-g, -f) = \frac{7}{2}, \frac{1}{2}$

$$\text{Radius of circle} = \sqrt{g^2 + f^2 - c} \quad = \sqrt{\left[-\frac{7}{2}\right]^2 + \left[-\frac{1}{2}\right]^2}$$

$$= \frac{1}{2}\sqrt{50}$$

$$= 3.54$$

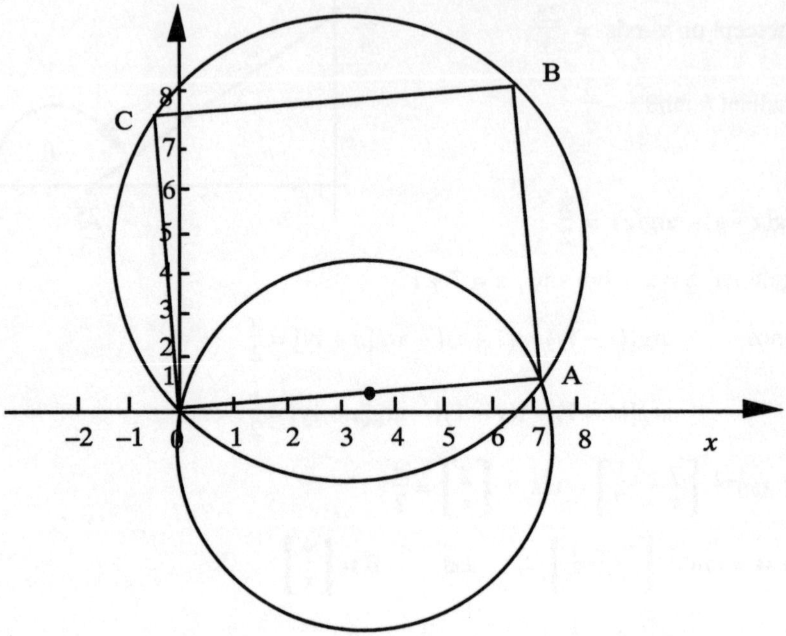

Question 3

a) Express the complex numbers 1 + i and 1 – i in trigonometric form.

Find the modulus and argument of $\dfrac{(1+i)^5}{(1-i)^7}$

[The argument θ of a complex number is measured in radians and is such that $-\pi < \theta \leq \pi$.]

b) Sketch the locus of the point in the Argand diagram representing the complex number z, where

$$|z - 1 - i| = \sqrt{2}\,.$$

c) Given that α and β are real numbers, show that

$$e^{i\alpha} + e^{i\beta} = [2\cos\tfrac{1}{2}(\alpha - \beta)]e^{i\tau}, \quad \text{where } \tau = \tfrac{1}{2}(\alpha + \beta).$$

Cambridge June 1986

Suggested solution

a) In general $x + iy = r\{\cos\theta + i\sin\theta\}$

where $r = \sqrt{x^2 + y2}$ and $\theta = \tan^{-1}\dfrac{y}{x}$

hence $1 + i \Rightarrow \quad x = 1$ and $y = 1$

ie $r = \sqrt{1^2 + 1^2} = \sqrt{2}$ and $\theta = \tan^{-1} 1$

$$\theta = \frac{\pi}{4}, \frac{5\pi}{4}, \frac{9\pi}{4} \text{ etc}$$

in general $\theta = n\pi + \dfrac{\pi}{4}$

hence $1 + i = \sqrt{2}\{\cos(n\pi + \pi/4) + i\sin(n\pi + \pi/4)\}$

For $1 - i \Rightarrow \quad x = 1$ and $y = -1$

ie $r = \sqrt{1^2 + (-1)^2} = \sqrt{2}$ and $\theta = \tan^{-1}(-1)$

$$\theta = \frac{3\pi}{4}, \frac{7\pi}{4}, \frac{11\pi}{4} \text{ etc}$$

In general $\theta = n\pi + \dfrac{3\pi}{4}$

hence $1 - i = \sqrt{2}\left\{\cos\left(n\pi + \dfrac{3\pi}{4}\right) + i\sin\left(n\pi + \dfrac{3\pi}{4}\right)\right\}$

$\dfrac{(1+i)^5}{(1-i)^7}$ ..... multiply the numerator and denominator by the conjugate of the denominator

$$= \frac{(1+i)^5(1+i)^7}{(1-i)^7(1+i)^7} = \frac{(1+i)^{12}}{[1-(-1)]^7} = \frac{(1+i)^{12}}{2^7}$$

Apply de Moivre's theorem

$$(1+i)^{12} = (\sqrt{2})^{12}\left\{\cos 12\left(n\pi + \frac{\pi}{4}\right) + i\sin 12\left(n\pi + \frac{\pi}{4}\right)\right\}$$

hence $\dfrac{(1+i)^5}{(1-i)^7} = \dfrac{2^6}{2^7}\left\{\cos 12\left(n\pi + \dfrac{\pi}{4}\right) + i\sin 12\left(n\pi + \dfrac{\pi}{4}\right)\right\}$

thus Modulus $= \dfrac{2^6}{2^7} = \dfrac{1}{2}$

and Argument $= 12\left(\dfrac{\pi}{4}\right)$ putting $n = 0$

$= 3\pi$

but Argument has limits $-\pi < \theta \le \pi$

This is satisfied if $\dfrac{-\pi}{3} < \theta \le \dfrac{\pi}{3}$

b) $|z - 1 - i| = \sqrt{2}$ where $z = x + iy$

$|z - 1 - i| = |x + iy - 1 - i| = |(x-1) + i(y-1)|$

$= \sqrt{(x-1)^2 + (y-1)^2}$

ie $\sqrt{(x-1)^2 + (y-1)^2} = \sqrt{2}$

square $x^2 - 2x + 1 + y^2 - 2y + 1 = 2$

$x^2 + y^2 - 2x - 2y = 0$

This can be compared to the general equation of a circle

$$x^2 + y^2 + 2gx + 2fy + c = 0$$

ie $2g = -2$ $\quad g = -1$

$2f = -2$ $\quad f = -1$

centre of circle is $(-g, -f) \Rightarrow (1, 1)$

radius of circle $= \sqrt{g^2 + f^2 - c} = \sqrt{1^2 + 1^2 - 0} = \sqrt{2}$

required locus is a circle, centre (1, 1), radius $\sqrt{2}$

c) To show $e^{i\alpha} + e^{i\beta} = 2\cos\frac{1}{2}(\alpha - \beta).e^{i\tau}$ where $\tau = \frac{1}{2}(\alpha + \beta)$

now $e^{i\theta} = \cos\theta + i\sin\theta$

hence $e^{i\alpha} + e^{i\beta} = \cos\alpha + i\sin\alpha + \cos\beta + i\sin\beta$

$= 2\cos\frac{1}{2}(\alpha + \beta).\cos\frac{1}{2}(\alpha - \beta) + i\left\{2\sin\frac{1}{2}(\alpha + \beta).\cos\frac{1}{2}(\alpha - \beta)\right\}$

$= 2\left\{\cos\frac{1}{2}(\alpha + \beta) + i\sin\frac{1}{2}(\alpha + \beta)\right\}\cos\frac{1}{2}(\alpha - \beta)$

$= 2\cos\frac{1}{2}(\alpha - \beta)\ \{\cos\tau + i\sin\tau\}$

$= 2\cos\frac{1}{2}(\alpha - \beta).e^{i\tau}$

Question 4

a) Given that $w = \dfrac{3 - 2i}{4 - i}$, express w in the form x + iy, and find | w | and arg w.

b) Sketch in an Argand diagram the set of points representing all complex numbers z satisfying both of the inequalities:

$$|z - 2i| \le 2 \quad \text{and} \quad |z - 2i| \le |z|.$$

c) Use de Moivre's theorem to express sin 5θ as a polynomial in sin θ.

Cambridge November 1987

Suggested solution

a) $w = \dfrac{3 - 2i}{4 - i}$

Multiply both the numerator and the denominator by the conjugate of the denominator ie by (4 + i)

$$w = \frac{(3 - 2i)(4 + i)}{(4 - i)(4 + i)} = \frac{12 - 5i - 2i^2}{4 - i^2}$$

$$= \frac{14 - 5i}{5} = \frac{14}{5} - i$$

$$|w| = \sqrt{\left[\frac{14}{5}\right]^2 + (-1)^2} = \sqrt{\frac{196 + 25}{25}} = 2.97$$

$$\arg w = \tan^{-1}\frac{-1}{(14/5)} = \tan^{-1}\frac{-5}{14} = -0.3430 \text{ radian}$$

b) $$|z - 2i| \le 2$$

z is a complex number of form x + iy

$$|x + iy - 2i| \le 2$$

$$|x + i(y - 2)| \le 2$$

ie $$\sqrt{x^2 + (y - 2)^2} \le 2$$

$$x^2 + (y - 2)^2 \le 4$$

$$x^2 + y^2 - 4y + 4 - 4 \le 0$$

$$x^2 + y^2 - 4y \le 0$$

compare this expression with the standard equation for a circle

$x^2 + y^2 + 2gx + 2fy + c = 0$ where the centre is given by (–g, –f) and radius = $\sqrt{(g^2 + f^2 - c)}$.

ie $x^2 + y^2 - 4y - 0$ represents a circle with centre (0, +2), radius = 2, inequality sign indicates the region within the circle.

also
$$|z - 2i| \leq |z|$$
$$|x + iy - 2i| \leq |x + iy|$$
$$|x + i(y - 2)| \leq |x + iy|$$

ie
$$\sqrt{x^2 + (y - 2)^2} \leq \sqrt{x^2 + y^2}$$
$$x^2 + y^2 - 4y + 4 \leq x^2 + y^2$$
$$4 \leq 4y$$
$$1 \leq y$$

ie $1 = y$ represents a straight line parallel to the x-axis, inequality sign indicates the region lies above the line.

Combining the two results the Argand diagram is as:

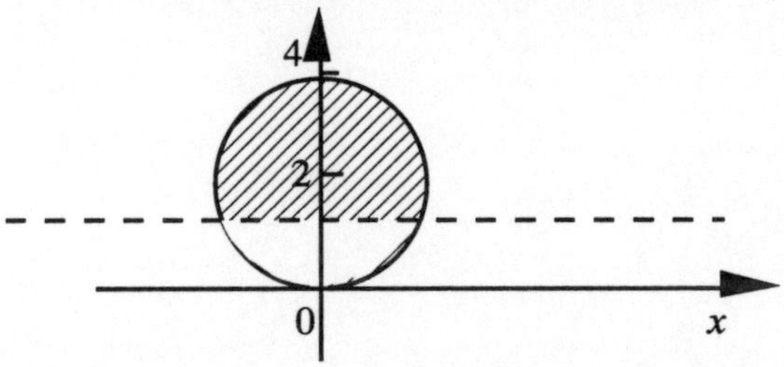

The required set of points lies within the shaded area.

c) Consider $(\cos\theta + i \sin\theta)^5$

expand using Pascall's triangle to find coefficients

(cf key formulae – chapter 2, page 19).

$\Rightarrow \cos^5\theta + 5 \cos^4\theta(i \sin\theta) + 10 \cos^3\theta(i \sin\theta)^2$

$+ 10 \cos^2\theta(i \sin\theta)^3 + 5 \cos\theta(i \sin\theta)^4 + (i \sin\theta)^5$

$\Rightarrow [\cos^5\theta - 10 \cos^3\theta.\sin^2\theta + 5 \cos\theta.\sin^4\theta]$

$+ i [5 \cos^4\theta.\sin\theta - 10 \cos^2\theta.\sin^3\theta + \sin^5\theta]$ ..... (i)

Using de Moivre's theorem

$(\cos\theta + i \sin\theta)^5 = \cos(5\theta) + i \sin(5\theta)$ ..... (ii)

equating like parts of (i) and (ii)

$$\sin(5\theta) = 5\cos^4\theta.\sin\theta - 10\cos^2\theta.\sin^3\theta + \sin^5\theta$$
$$= 5(1 - \sin^2\theta)^2.\sin\theta - 10(1 - \sin^2\theta).\sin^3\theta + \sin^5\theta$$
$$= 5(1 - 2\sin^2\theta + \sin^4\theta)\sin\theta - 10\sin^3\theta + 10\sin^5\theta + \sin^5\theta$$

ie $\sin(5\theta) = 16\sin^5\theta - 20\sin^3\theta + 5\sin\theta$

10 VECTORS

10.1 Key formulae and ideas

1. Scalar quantity ..... has magnitude only

   Vector quantity ..... has magnitude and direction and can be represented by a straight line

   where length ⇒ magnitude

   direction of line ⇒ direction of vector

   Vectors may be written $\underline{a}$; $\overrightarrow{OA}$; $\underline{OA}$; **a** or **OA** in BOLD type.

   If $\underline{i}, \underline{j}, \underline{k}$ are unit vectors in three mutually perpendicular directions then

   vector $\underline{r} = a\underline{i} + b\underline{j} + c\underline{k}$ and $|\underline{r}| = \sqrt{a^2 + b^2 + c^2}$ a, b, c are scalars.

2. Addition of vectors is by the triangle rule

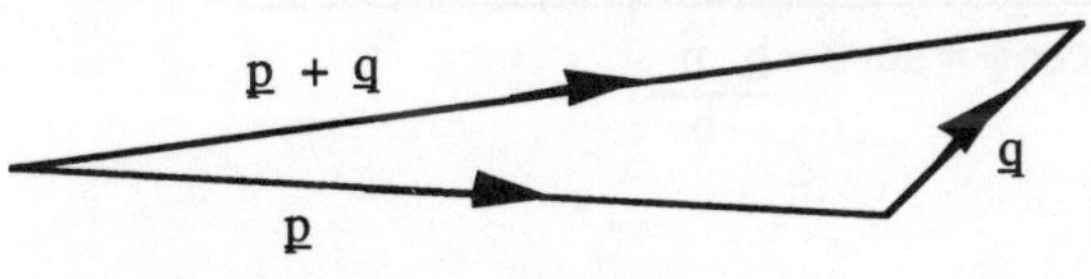

3. Vector equation of a straight line

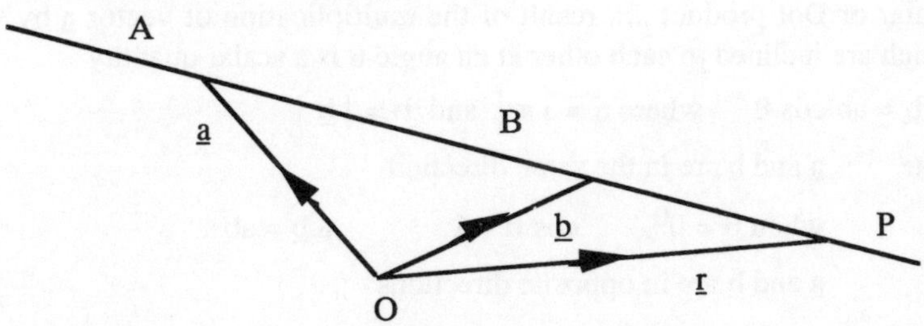

$\underline{r} = \underline{a} + \mu(\underline{b} - \underline{a})$ where μ is a scalar.

4. Perpendicular distance from a point P to a line where A is a position on the line.

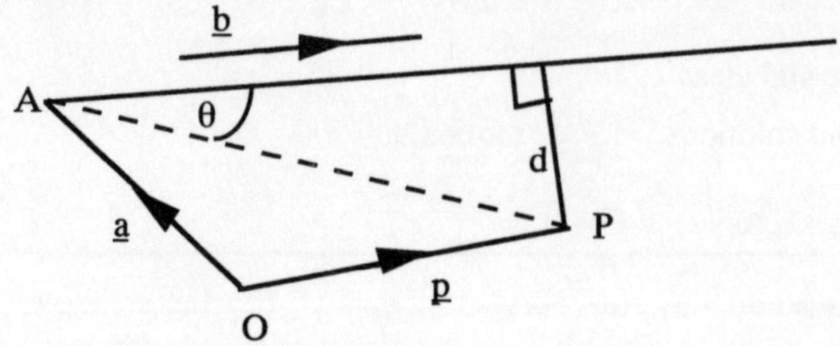

$\underline{b}$ = vector of line

$\underline{a}$ = position vector of A

$\underline{p}$ = position vector of P

Perpendicular distance = d = $\left| \dfrac{\underline{b} \times (\underline{p} - \underline{a})}{b} \right|$

5. Acute angle between a line and a plane.

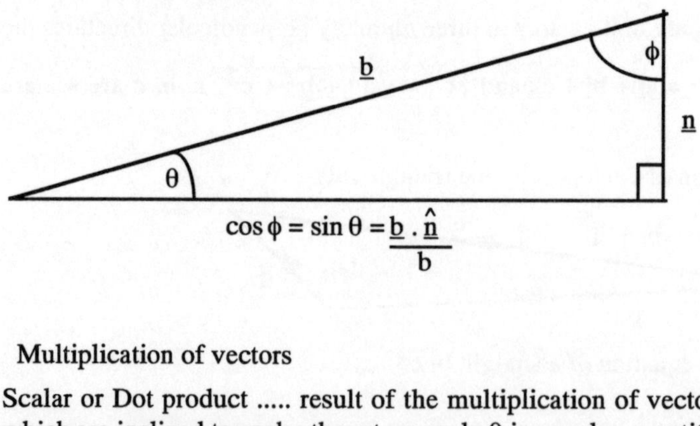

$$\cos\phi = \sin\theta = \frac{\underline{b} \cdot \hat{\underline{n}}}{b}$$

6. Multiplication of vectors

i) Scalar or Dot product ..... result of the multiplication of vector $\underline{a}$ by vector $\underline{b}$ which are inclined to each other at an angle θ is a scalar quantity

$\underline{a} \cdot \underline{b} = ab\cos\theta$ where $a = |\underline{a}|$ and $b = |\underline{b}|$

Note $\underline{a}$ and $\underline{b}$ are in the same direction

when $\theta = 0^0$, $\cos\theta = 1$, $\underline{a}.\underline{b} = ab$

$\underline{a}$ and $\underline{b}$ are in opposite directions

when $\theta = 180^0$, $\cos\theta = -1$, $\underline{a}.\underline{b} = -ab$

$\underline{a}$ and $\underline{b}$ are perpendicular

when $\theta = 90^0$, $\cos\theta = 0$, $\underline{a}.\underline{b} = 0$

In component form $\underline{i}.\underline{i} = i^2 = 1$

$\underline{i}.\underline{j} = i.j.\cos 90^0 = 0$

Angle between vectors $\cos\theta = \dfrac{a.b}{ab}$

ii) Vector (cross) product ..... of two vectors $\underline{a}$ and $\underline{b}$ bearing an angle θ to each other is a vector of magnitude ab sin θ in a direction perpendicular to the plane containing $\underline{a}$ and $\underline{b}$

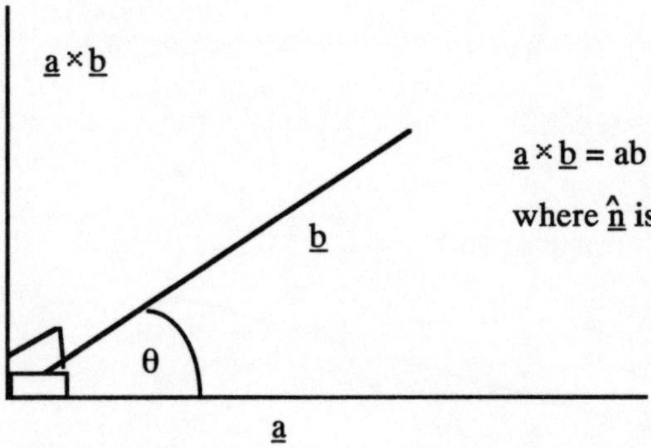

$\underline{a} \times \underline{b} = ab \sin\theta.\hat{\underline{n}}$

where $\hat{\underline{n}}$ is a unit vector

parallel vectors $\sin\theta = 0$ $\underline{a} \times \underline{b} = 0$

perpendicular vectors $\sin\theta = 90^o$

$|\,\underline{a} \times \underline{b}\,| = ab \sin 90^o = ab$

$\underline{a} \times \underline{b} = ab.\underline{n}$

In component form

$\underline{i} \times \underline{j} = \underline{k}$

$\underline{j} \times \underline{k} = \underline{i}$ but $\underline{i} \times \underline{i} = \underline{j} \times \underline{j} = \underline{k} \times \underline{k} = 0$

$\underline{k} \times \underline{i} = \underline{j}$

7. **Standard equation of a plane defined in terms of a unit vector perpendicular to the plane ... $\underline{r}\,\hat{\underline{n}} = d$**

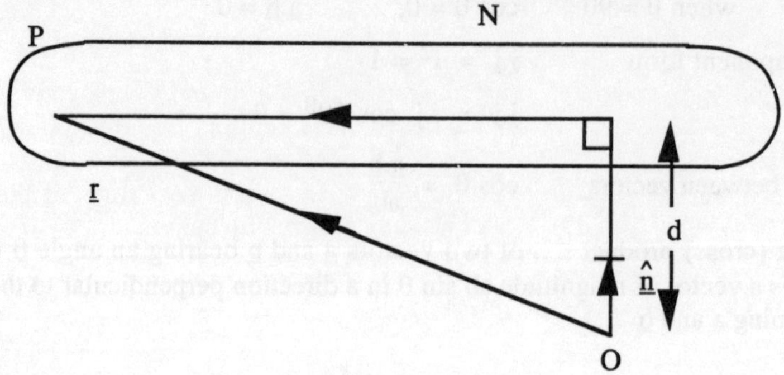

10.2 Questions and solutions

Question 1

The points A, B and C in space have position vectors a, b and c, respectively, with respect to the origin O. The vectors **a** and **b** have lengths of 1 unit and 2 units, respectively, and the angle AOB is 60°. The vector **c** has length 3 units and its direction is perpendicular to the plane AOB. Write down the values of the scalar products **a.b**, **b.c**, **b.b** and **a.c**. Calculate the lengths of the vectors **b** – **c** and **a** – **b**, and the angle between the lines BC and AB.

The line OB is produced to D so that BD has length 1 unit. The point E is the mid point of AB and the line OE meets AD at F. By writing the position vector of F in the form

$$\frac{2p\mathbf{a} + 3q\mathbf{b}}{2p + 2q},$$

or otherwise, find the position vector of F.

Oxford June 1986

Suggested solution

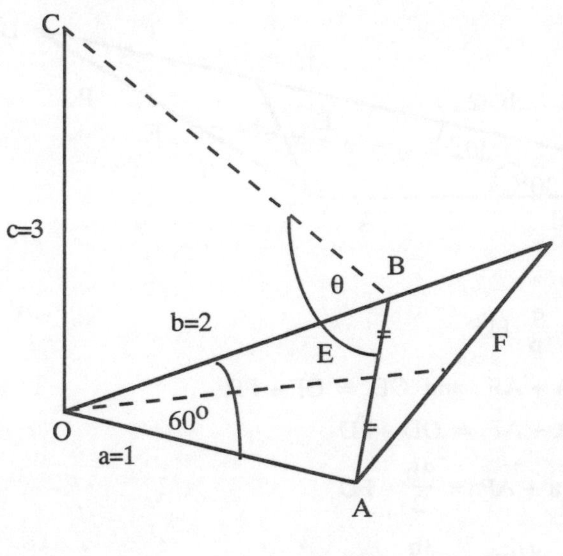

Scalar products

$\mathbf{a.b} = 1 \times 2 \times \cos 60^o \quad = 1$

$\mathbf{b.c} = 2 \times 3 \times \cos 90^o \quad = 0$

$\mathbf{b.b} = 2 \times 2 \times \cos 0^o \quad = 4$

$\mathbf{a.c} = 1 \times 3 \times \cos 90^o \quad = 0$

length of vector

$\mathbf{b} - \mathbf{c} = \sqrt{3^2 + 2^2} \quad = \sqrt{13}$

$\mathbf{a} - \mathbf{b} = \sqrt{2^2 + 1^2} \quad = \sqrt{5}$

angle CBA

$\cos\theta \quad = \quad \dfrac{\text{vectors: } \underline{(\mathbf{b}-\mathbf{c}).(\mathbf{a}-\mathbf{b})}}{\text{scalars: } (\mathbf{b}-\mathbf{c}).(\mathbf{a}-\mathbf{b})}$

$$= \frac{\mathbf{b.a} - \mathbf{b.b} - \mathbf{c.a} + \mathbf{c.b}}{\sqrt{13} \times \sqrt{5}} = \frac{1 - 4 - 0 + 0}{\sqrt{13} \times \sqrt{5}} = -0.3721$$

$\theta \quad = \quad 111.8^o$

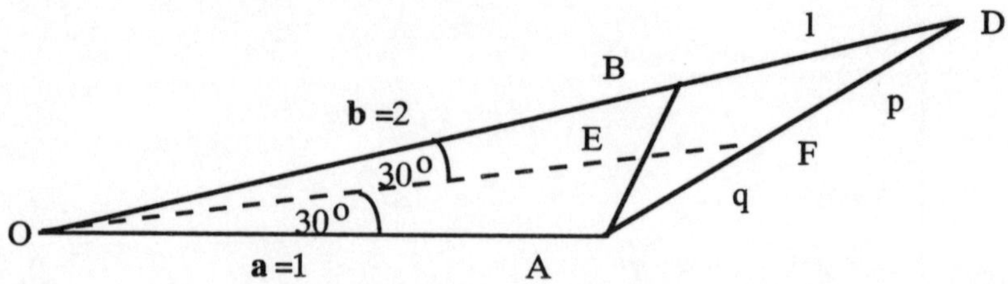

Let $\dfrac{AF}{FD} = \dfrac{q}{p}$ ie $AF = \dfrac{q}{p}\ FD$

from diagram $\quad OF = OA + AF \quad$ and $\quad OD = OF + FD$

hence $\quad OF = OA + AF = OD - FD$

$$\mathbf{a} + AF = \frac{3\mathbf{b}}{2} - FD$$

substitute for AF $\quad \mathbf{a} + \dfrac{q}{p} FD = \dfrac{3\mathbf{b}}{2} - FD$

hence $$FD\left[\frac{q+p}{p}\right] = \frac{3\mathbf{b}}{2} - \mathbf{a}$$

$$FD = \frac{(3\mathbf{b} - 2\mathbf{a})}{2} \cdot \frac{p}{(p+q)}$$

$$AF = \frac{q}{p} \cdot \frac{(3\mathbf{b} - 2\mathbf{a})}{2} \cdot \frac{p}{(p+q)}$$

$$= \frac{(3\mathbf{b} - 2\mathbf{a})}{2} \cdot \frac{q}{(p+q)}$$

thus $$OF = OA + AF = \mathbf{a} + \frac{(3\mathbf{b} - 2\mathbf{a})}{2} \cdot \frac{q}{(p+q)}$$

$$= \frac{2\mathbf{a}(p+q) + (3\mathbf{b} - 2\mathbf{a})q}{2(p+q)}$$

$$OF = \frac{2\mathbf{a}p + 3\mathbf{b}q}{2(p+q)}$$

From triangle OFD

$$\frac{3}{\sin O\hat{F}D} = \frac{p}{\sin 30^o} \qquad \text{and} \qquad \frac{1}{\sin O\hat{F}A} = \frac{q}{\sin 30^o}$$

$$\sin O\hat{F}D = \sin O\hat{F}A = \frac{3 \times \frac{1}{2}}{p} = \frac{1 \times \frac{1}{2}}{q}$$

$$\frac{q}{p} = \frac{1}{3}$$

Also $OD = OA + AD$ and $AD = q + p$

$3 = 1 + (p + q)$ ie $p + q = 2$

hence $p = \frac{3}{2}$ and $q = \frac{1}{2}$

position vector of F = OF $$= \frac{2(1)(\frac{3}{2}) + 3(2)(\frac{1}{2})}{2(2)}$$

$$= \frac{3}{2}$$

Question 2

The point A has position vector $\begin{bmatrix}1\\2\\3\end{bmatrix}$, the line l has equation

$$r = \begin{bmatrix}1\\4\\-1\end{bmatrix} + t\begin{bmatrix}1\\-1\\1\end{bmatrix}$$

and the plane p has equation $r \cdot \begin{bmatrix}-1\\2\\2\end{bmatrix} = 2$

i) Find the point of intersection of l and p

ii) Find the length of the perpendicular from A to p

iii) Find the length of the perpendicular from A to l

iv) Find the acute angle between l and p, correct to the nearest 0.1^o.

Cambridge June 1988

Suggested solution

i) Intersection of line l with plane p:

rearrange the equation of the line l

$$r = (1 + t)i + (4 - t)j + (-1 + t)k$$

substitute into the equation of the plane

$$[(1 + t)i + (4 - t)j + (-1 + t)k].[-i + 2j + 2k] = 2$$

$$(1 + t)(-1) + (4 - t)(2) + (-1 + t)(2) = 2$$

at point of intersection $t = 3$

ie coordinates are $x = (1 + t) = 4$

$y = (4 - t) = 1$

$z = (-1 + t) = 2$

ii) Length of perpendicular from A to the plane

put $r \cdot \begin{bmatrix}-1\\2\\2\end{bmatrix} = 2$ into the standard form $\mathbf{r}.\hat{\mathbf{n}} = d$ by dividing by n

$$n = |\,\mathbf{n}\,| = \sqrt{(-1)^2 + 2^2 + 2^2} = 3$$

ie $\quad \mathbf{r} \cdot \frac{(-i + 2j + 2k)}{3} = \frac{2}{3}$

$$\mathbf{r}.\hat{\mathbf{n}} = \frac{2}{3}$$

This plane is $\left(\frac{2}{3}\right)$ units from the origin.

Now imagine another plane through A, position vector = i + 2j + 3k

using equation for a plane $\quad r.\hat{n} = a.\hat{n}$

$$\mathbf{r} \cdot \frac{(-i + 2j + 2k)}{3} = (i + 2j + 3k) \cdot \frac{(-i + 2j + 3k)}{3}$$

$$= \frac{(1)(-1) + (2)(2) + (3)(3)}{3}$$

$$= 4$$

This imaginary plane through A is 4 units from the origin

$\therefore$ point A is $4 - \frac{2}{3} = \frac{10}{3}$ units from the plane.

iii) Length of perpendicular from A to line ***l***

distance of point from a line = d = $\left| \frac{\mathbf{b} \times (\mathbf{p} - \mathbf{a})}{b} \right|$

in question

p = 1 + 2j + 3k ..... position of A (which is the point P in formula (4) on page 136 ... unfortunate choice of symbol in question)

a = i + 4j – k

b = i – j + k

hence $\quad (\mathbf{p} - \mathbf{a}) = -2j + 3k$

and $\quad \mathbf{b} \times (\mathbf{p} - \mathbf{a}) = (i - j + k) \times (-2j + 4)$

$$= [(-1)(4) - (1)(-2)]i + [(1)(0) - (1)(4)]j + [(1)(-2) - (-1)(0)]k$$

$$= -2i - 4j - 2k$$

$\therefore \quad |\mathbf{b} \times (\mathbf{p} - \mathbf{a})| = \sqrt{(-2)^2 + (-4)^2 + (-2)^2} = \sqrt{24}$

and $\quad b = |\mathbf{b}| = \sqrt{1^2 + (-1)^2 + 1^2} = \sqrt{3}$

hence required distance $\quad = \frac{\sqrt{24}}{\sqrt{3}} = \sqrt{8} = 2\sqrt{2}$

iv) Acute angle between *l* and *p*

this is obtained by the scalar product using the direction component of the line *l* and the plane of formula (6) on page 136

line *l* $\quad \mathbf{r} = i + 4j - k + t(i - j + k)$

direction component $\quad \mathbf{b} = (i - j + k)$

$$b = \sqrt{1^2 + (-1)^2 + 1^2} = \sqrt{3}$$

plane p $\quad \mathbf{r}.(-i + 2j + 2k) = 2$

unit vector

$$\hat{\mathbf{n}} = \frac{-i + 2j + 2k}{\sqrt{(-1)^2 + 2^2 + 2^2}}$$

$$= \frac{-i + 2j + 2k}{3}$$

hence

$$\sin\theta = \frac{(i - j + k).\frac{(-i + 2j + 2k)}{3}}{\sqrt{3}}$$

$$= \frac{(1)(-1) + (-1)(2) + (1)(2)}{3\sqrt{3}}$$

$$= -\frac{1}{3\sqrt{3}}$$

Angle between line *l* and plane p = – 11.09 = – 11.1°

11 DIFFERENTIAL EQUATIONS

11.1 Key formulae and ideas

1. Equations of First Order and First Degree

equations of the form $M + N\frac{dy}{dx} = 0$

ie $M\,dx + N\,dy = 0$

It is not possible to solve a 'general' equation in terms of a finite number of known functions. There are, however, a number of special types.

a) Exact equations

eg $y.dx + x.dy = 0$ ..... $d(xy) = 0$ ..... integrating $xy = c$

b) Equations solvable by separation of variables

eg $\frac{dx}{x} = \cos y.dy$ ..... integrating $\log_e x = \sin y + c$

c) Homogeneous equations

eg of the form $\frac{dy}{dx} = f\left[\frac{y}{x}\right]$

To test whether a function of x and y can be written in the form of the right hand side, put $y = v.x$ ..... if the result is that all the x's cancel the test is successful.

eg if $\frac{dy}{dx} = \frac{x^2 + y^2}{2x^2}$ ..... put $y = v.x$ and $\frac{dy}{dx} = v + x\,.\,\frac{dv}{dx}$

ie $v + x\,.\,\frac{dv}{dx} = \frac{1 + v^2}{2}$ ..... separate variables to continue

d) Linear equations of the First Order

eg $\frac{dy}{dx} + Py = Q$ ..... where P and Q are functions of x but not of y eg if $\frac{dy}{dx} + \frac{1}{x}(y) = x^2$

ie $x\frac{dy}{dx} + y = x^3$ ..... $\frac{d}{dx}(xy) = x^3$

$xy = \frac{1}{4}x^4 + k$

In this case x was an obvious integrating factor.

$e^{\int P dx}$ is a general integrating factor to use in this type of equation (multipying both sides).

2. Linear equations with Constant Coefficients of the form

$$p_0 \frac{d^n y}{dx^n} + p_1 \frac{d^{n-1} y}{dx^{n-1}} + \ldots\ldots p_n y = f(x)$$

These are important equations in the study of mechanical, acoustical and electrical vibrations.

a) Equations of the First order ..... simplest case when n = 1 and $f(x) = 0$

ie $p_0 \frac{dy}{dx} + p_1 y = 0 \Rightarrow y = Ae^{-(p_1 x/p_0)}$

b) Equations of the Second order

The solution consists of two parts

i) the Complementary Function ..... where n = 2 and $f(x) = 0$

ie $p_0 \frac{d^2y}{dx^2} + p_1 \frac{dy}{dx} + p_2 y = 0$

and which is solved by using an auxiliary quadratic equation.

Section (a) implies $y = A\,e^{mx}$ where m is a constant.

Differentiating twice and substituting leads to

$A\,e^{mx}(p_0 m^2 + p_1 m + p_2) = 0$

Hence, if m is a root of $p_0 m^2 + p_1 m + p_2 = 0$

then $y = A\,e^{mx}$ is a solution whatever the value of A.

$p_0 m^2 + p_1 m + p_2 = 0$ is known as the 'auxiliary equation'.

If $(p_1)^2 > 4(p_0)(p_2)$ try $y = A\,e^{mx} + B\,e^{nx}$

$(p_1)^2 = 4(p_0)(p_2)$ try $y = (A + Bx)\,e^{mx}$

$(p_1)^2 < 4(p_0)(p_2)$ try $y = e^{mx} \cos(Bx + \theta)$

ii) the Particular Integral

This depends on the nature of $f(x)$ and the solution is largely trial and error within certain guidelines

eg form of $f(x)$:	trial solution:
e^{mx}	$A e^{mx}$
a constant	A
$px + q$	$mx + k$
$p \sin x$ $p \cos x$ $p \sin x + q \cos x$	] $A \sin x + B \cos x$

11.2 Questions and solutions

Question 1

a) Solve the differential equations

i) $\dfrac{d^2y}{dx^2} - 4y = 0$

ii) $\dfrac{d^2y}{dx^2} - 4x = 0$

given that in each case y = 2 and $\dfrac{dy}{dx} = 1$ when $x = 0$.

b) Express $f(x) = \dfrac{x^2 + 7x + 2}{(1 + x^2)(2 - x)}$ in partial fractions

Hence prove that $\displaystyle\int_0^1 f(x)\, dx = \frac{11}{2} \ln 2 - \frac{1}{4}\pi$.

Oxford June 1988

Suggested solution

a) i) $\dfrac{d^2y}{dx^2} - 4y = 0$

is of the form $p_0 \dfrac{d^2y}{dx^2} + p_1 \dfrac{dy}{dx} + p_2 y = 0$

Try $y = A e^{mx}$ ie $\dfrac{dy}{dx} = Am e^{mx}$ and $\dfrac{d^2y}{dx^2} = Am^2 e^{mx}$

substitute into original differential equation

$A e^{mx} (m^2 - 4) = 0$ hence $m^2 - 4 = 0$

$m = \pm 2$

general solution $\Rightarrow$ $y = A e^{2x} + B e^{-2x}$

to find values of A and B

when $x = 0$ $y = 2 = A e^0 + B e^0$ ie $2 = A + B$

$x = 0$ $\dfrac{dy}{dx} = 1 = 2A e^0 - 2B e^0$ ie $1 = 2A - 2B$

solving for A and B $\quad A = \frac{5}{4}$ and $B = \frac{3}{4}$

hence general solution is $y = \frac{5}{4}e^{2x} + \frac{3}{4}e^{-2x}$

ii) $\frac{d^2y}{dx^2} - 4x = 0$ ie $\frac{d^2y}{dx^2} = 4x$

This can be integrated directly $\quad \frac{dy}{dx} = 2x^2 + k$

when $x = 0$ $\quad \frac{dy}{dx} = 1 = 0 + k$ ie $k = 1$

hence $\quad \frac{dy}{dx} = 2x^2 + 1$

Integrate a second time $\quad y = \frac{2}{3}x^3 + x + K$

when $x = 0$ $\quad y = 2 = 0 + 0 + K$ ie $K = 2$

hence $\quad y = \frac{2}{3}x^3 + x + 2$

b) Let $\frac{x^2 + 7x + 2}{(1 + x^2)(2 - x)} \Rightarrow \frac{A}{(2 - x)} + \frac{Bx + C}{(1 + x^2)}$

clear the denominators

$$x^2 + 7x + 2 = A(1 + x^2) + (Bx + C)(2 - x)$$

put $x = 2$ $\quad 4 + 14 + 2 = A(1 + 4) + (2B + C)(0)$

ie $\quad A = 4$

put $x = 0$ $\quad 2 = A + 2C$ $\quad$ ie $C = -1$

put $x = 1$ $\quad 10 = 8 + B - 1$ $\quad$ ie $B = 3$

hence $\quad \frac{x^2 + 7x + 2}{(1 + x^2)(2 - x)} \Rightarrow \frac{4}{(2 - x)} + \frac{(3x - 1)}{(1 + x^2)}$

so $\quad \int_0^1 \frac{x^2 + 7x + 2}{(1 + x^2)(2 - x)}\,dx = \int_0^1 \left[\frac{4}{(2 - x)} + \frac{(3x - 1)}{(1 + x^2)}\right] dx$

$$\int_0^1 \frac{4}{(2-x)}\,dx = -4\,[\log_e(2-x)]_0^1$$

$$= -4[\log_e(2-1) - \log_e(2-0)] = +4\log_e 2$$

$$\int_0^1 \frac{(3x-1)}{(1+x^2)}\,dx = \int_0^1 \frac{3x}{(1+x^2)}\,dx - \int_0^1 \frac{1}{(1+x^2)}\,dx$$

$$\int_0^1 \frac{3x}{(1+x^2)}\,dx = \frac{3}{2}\int_0^1 \frac{2x}{(1+x^2)}\,dx = \frac{3}{2}\,[\log_e(1+x^2)]_0^1$$

$$= \frac{3}{2}\,[\log_e 2 - \log_e 1\,] = \frac{3}{2}\log_e 2$$

$$\int_0^1 \frac{dx}{(1+x^2)}\,dx = [\tan^{-1}x]_0^1 = \tan^{-1}(1) - \tan^{-1}(0)$$

$$= \tan^{-1}(1) = \frac{\pi}{4}$$

hence $\int_0^1 f(x)\,dx = 4\log_e 2 + \frac{3}{2}\log_e 3 - \frac{\pi}{4}$

$$= \frac{11}{2}\log_e 2 - \frac{\pi}{4}$$

Question 2

a) Find the general solution of the differential equation

$$\frac{d^2y}{dx^2} + 4\frac{dy}{dx} + 13y = 25\,e^{2x}$$

b) Substitute $y = \frac{z}{x}$ in the differential equation

$$\frac{dy}{dx} = \frac{2xy + 1 - 2y}{x(x + 2)}$$

to obtain a differential equation relating z and x.

Solve this equation and hence find the general solution for y in terms of x.

Cambridge June 1987

Suggested solution

a) The general solution will consist of the sum of a Complementary Function and a Particular Integral.

Complementary function

Put $$\frac{d^2y}{dx^2} + 4\frac{dy}{dx} + 13y = 0$$

This implies a solution of the form $y = A\,e^{mx}$

ie $\frac{dy}{dx} = Am\,e^{mx}$ and $\frac{d^2y}{dx^2} = Am^2\,e^{mx}$

hence, substituting $A\,e^{mx}(m^2 + 4m + 13) = 0$

solving the auxiliary equation $m = \frac{-4 \pm \sqrt{4^2 - 4(1)(13)}}{2(1)}$

$$= \frac{-4 \pm \sqrt{-36}}{2}$$

$$m = -2 \pm 3i$$

Complementary function $\Rightarrow$ $y = A\,e^{(-2+3i)x} + B\,e^{(-2-3i)x}$

To modify the solution to present it without imaginary quantities

recall $e^{iqx} = \cos qx + i \sin qx$

$e^{-iqx} = \cos qx - i \sin qx$

thus $y = e^{-2x}\,[A(\cos 3x + i \sin 3x) + B(\cos 3x - i \sin 3x)]$

put $E = A + B$ and $F = i(A - B)$

where E and F are arbitrary constants just as A and B are

$$y = e^{-2x}\,[E \cos 3x + F \sin 3x]$$

let $E = C.\cos \alpha$ and $F = C.\sin \alpha$

where $C = \sqrt{E^2 + F^2}$ and $\tan \alpha = \dfrac{F}{E}$

thus $y = e^{-2x}\,[\cos 3x.C.\cos \alpha + \sin 3x.C.\sin \alpha]$

$$y = C.e^{-2x} \cos(3x - \alpha)$$

Particular integral

The right hand side of the original expression $= 25\, e^{2x}$

Try a solution of the form $y = G\, e^{2x}$

ie $\dfrac{dy}{dx} = 2G\, e^{2x}$ and $\dfrac{d^2y}{dx^2} = 4G\, e^{2x}$

substitute in original expression

$$4G\, e^{2x} + 8G\, e^{2x} + 13G\, e^{2x} = 25\, e^{2x}$$

ie $4G + 8G + 13G = 25 \quad \therefore\ G = 1$

Particular integral $y = e^{2x}$

General solution: $y = e^{2x} + C.e^{-2x} \cos(3x - \alpha)$

b) $$\frac{dy}{dx} = \frac{2xy + 1 - 2y}{x(x + 2)}$$

put $y = \dfrac{z}{x}$ thus $\dfrac{dy}{dx} = -\dfrac{z}{x^2} + \dfrac{1}{x} \cdot \dfrac{dz}{dx}$

so $$-\frac{z}{x^2} + \frac{1}{x} \cdot \frac{dz}{dx} = \frac{2z + 1 - 2(z/x)}{x(x + 2)}$$

ie $$\frac{dz}{dx} = \frac{2zx + x - 2z}{x(x + 2)} + \frac{z}{x}$$

$$= \frac{2zx + x - 2z + z(x + 2)}{x(x + 2)}$$

$$\frac{dz}{dx} = \frac{(3z + 1)}{(x + 2)}$$

To solve, separate the variables

$$\frac{dz}{(3z + 1)} = \frac{dx}{(x + 2)}$$

and integrate $\frac{1}{3}\log_e(3z + 1) = \log_e(x + 2) + \log_e k$

where k is a constant of integration

$$\log_e(3z + 1)^{\left(\frac{1}{3}\right)} = \log_e(x + 2) + \log_e k$$

ie $$(3z + 1)^{\left(\frac{1}{3}\right)} = k(x + 2)$$

$$3z = k^3(x + 2)^3 - 1$$

put $z = xy$ $$3xy = k^3(x + 2)^3 - 1$$

put $A = k^3$ $$y = \frac{1}{3x}[A(x + 2)^3 - 1]$$

DETAILS FOR DESPATCH OF PUBLICATIONS

Please insert your full name below

Please insert below the style in which you would like the correspondence from the Publisher addressed to you

TITLE Mr, Miss etc.	INITIALS	SURNAME/FAMILY NAME

Address to which study material is to be sent (please ensure someone will be present to accept delivery of your Publications).

POSTAGE & PACKING

You are welcome to purchase study material from the Publisher at 200 Greyhound Road, London W14 9RY, during normal working hours.

If you wish to order by post this may be done direct from the Publisher. Postal charges are as follows:

UK - Orders over £30: no charge. Orders below £30: £2.50. Single paper (last exam only): 50p
OVERSEAS - See table below

The Publisher cannot accept responsibility in respect of postal delays or losses in the postal systems.

DESPATCH All cheques must be cleared before material is despatched.

SUMMARY OF ORDER

Date of order:

				£
Cost of publications ordered:				
UNITED KINGDOM:				
OVERSEAS:	TEXTS		Suggested Solutions (Last exam only)	
	One	Each Extra		
Eire	£4.00	£0.60	£1.00	
European Community	£9.00	£1.00	£1.00	
East Europe & North America	£10.50	£1.00	£1.00	
South East Asia	£12.00	£2.00	£1.50	
Australia/New Zealand	£13.50	£4.00	£1.50	
Other Countries (Africa, India etc)	£13.00	£3.00	£1.50	
Total cost of order:				£

Please ensure that you enclose a cheque or draft payable to THE HLT GROUP LTD for the above amount, or charge to ❑ Access ❑ Visa ❑ American Express

Card Number

Expiry Date .... Signature ....

ORDER FORM – 'A' LEVEL AND GCSE PUBLICATIONS

Please indicate on the form below the titles and quantity of each which you require. Then complete the despatch details overleaf where postal charges are set out.

'A' LEVEL TEXTBOOKS	Cost £	Quantity	Total £
Accounting	10.95		
Business Studies	10.95		
Economics	10.95		
Applied Mathematics	9.95		
Pure Mathematics	9.95		
Statistics	9.95		
General Principles of English Law	10.95		
Constitutional Law	9.95		
Sociology	9.95		
'A' LEVEL CASEBOOK			
General Principles of English Law	10.95		
GCSE			
Law Textbook	7.95		
Law Casebook	6.95		
		Total Cost £	

HLT PUBLICATIONS

The books listed above should be available from your local bookshop. In case of difficulty, however, they can be obtained direct from the publisher using this order form. Telephone, Fax or Telex orders will also be accepted. Quote your Access, Visa or American Express card numbers for priority orders. To order direct from publisher please enter cost of titles you require, fill in despatch details overleaf and send it with your remittance to The HLT Group Ltd.